Green Energy Systems

Green Energy Systems

Design, Modeling, Synthesis and Applications

Edited by

Vinod Kumar Singh

Naresh Bangari

Ratnesh Tiwari

Vikas Dubey

Akash Kumar Bhoi

Thanikanti Sudhakar Babu

ELSEVIER

ACADEMIC PRESS

An imprint of Elsevier

Academic Press is an imprint of Elsevier
125 London Wall, London EC2Y 5AS, United Kingdom
525 B Street, Suite 1650, San Diego, CA 92101, United States
50 Hampshire Street, 5th Floor, Cambridge, MA 02139, United States
The Boulevard, Langford Lane, Kidlington, Oxford OX5 1GB, United Kingdom

ISBN: 978-0-323-95108-1

For information on all Academic Press publications visit our website at
https://www.elsevier.com/books-and-journals

Publisher: Charlotte Cockle
Acquisitions Editor: Peter Adamson
Editorial Project Manager: Naomi Robertson
Production Project Manager: Sajana Devasi PK
Cover Designer: Greg Harris

Typeset by TNQ Technologies

Contents

List of contributors

Jacobo Aguillon-Garcia Korea Advanced Institute of Science and Technology, Daejeon, Korea

Ali Q. Al-Shetwi Electrical Engineering Department, Fahad Bin Sultan University, Tabuk, Saudi Arabia

Ameena Al-Sumaiti Advanced Power and Energy Center, Khalifa University, Abu Dhabi, United Arab Emirates

A. Arputhalatha Department of Physics, Arizona State University, Tempe, AZ, United States

Thanikanti Sudhakar Babu Department of Electrical and Electronics Engineering, Chaitanya Bharathi Institute of Technology (CBIT), Hyderabad, Telangana, India

Praveen Kumar Balachandran Department of Electrical and Electronics Engineering, Vardhaman College of Engineering, Hyderabad, Telangana, India

Harold R. Chamorro KTH, Royal Institute of Technology, Stockholm, Sweden

Ramesh Chandra Thakur Department of Chemistry, Himanchal Pradesh University, Shimla, Himachal Pradesh, India

Yogesh K. Chauhan Electrical Engineering Department, Kamla Nehru Institute of Technology, Sultanpur, Uttar Pradesh, India

Prince Winston David Department of Electrical and Electronics Engineering, Kamaraj College of Engineering and Technology, Virudhunagar, Tamil Nadu, India

Poobana Dharmalingam School of Biosciences and Technology, VIT University, Vellore, Tamil Nadu, India

Rajvikram Madurai Elavarasan Clean and Resilient Energy Systems Laboratory, Texas A&M University, Galveston, TX, United States

Shimaa M. Elsaeed Petroleum Applications Division, Egyptian Petroleum Research Institute (EPRI), Cairo, Egypt

S. Esther Gandhimathi Department of Electronics and Communication Engineering, Dhanalakshmi College of Engineering, Chennai, Tamil Nadu, India

Wilfredo C. Flores Universidad Tecnológica Centroamericana, Tegucigalpa, Honduras

K. Gomathi Department of Mechatronics Engineering, Kongu Engineering College, Perundurai, India

Francisco Gonzalez-Longatt University of South-Eastern Norway, Porsgrunn, Norway

G.V. Harshini Department of Mechanical Engineering, Amrita School of Engineering, Amrita Vishwa Vidyapeetham, Coimbatore, India

Bryan R. Jovel-Zelaya Universidad Tecnológica Centroamericana, Tegucigalpa, Honduras

Karthik Kannan BrainPool Program Postdoctoral Fellow, School of Advanced Materials Science and Engineering, Kumoh National Institute of Technology (KIT), Gyeongbuk, Republic of Korea

Charalambos Konstantinou Division of Computer, Electrical and Mathematical Sciences and Engineering, King Abdullah University of Science and Technology (KAUST), Thuwal, Saudi Arabia

Anurag Koushal Electrical Engineering Department, School of Engineering, Gautam Buddha University, Greater Noida, Uttar Pradesh, India

Murugaperumal Krishnamoorthy Department of Electrical and Electronics Engineering, Vardhaman College of Engineering, Hyderabad, Telangana, India

Ashish Kumar Department of Chemistry, Faculty of Technology and Sciences, Lovely Professional University, Phagwara, Punjab, India; NCE, Department of Science and Technology, Government of Bihar, Patna, India

N. Kumar Swamy Department of Physics, ISBM University, Nawapara (Kosmi), Gariyaband, Chhattisgarh, India

Soumitra Mukhopadhyay Doosan Power Systems India Pvt. Ltd., Kolkata, West Bengal, India

Palpandian Murugesan Department of Electrical and Electronics Engineering, SCAD College of Engineering and Technology, Cheranmahadevi, Tirunelveli, Tamil Nadu, India

Karuppiah Natarajan Department of Electrical and Electronics Engineering, Vardhaman College of Engineering, Hyderabad, Telangana, India

Nidhi Department of Chemistry, Faculty of Technology and Sciences, Lovely Professional University, Phagwara, Punjab, India

Rupendra Kumar Pachauri Electrical and Electronics Engineering Department, School of Engineering, University of Petroleum and Energy Studies, Dehradun, Uttarakhand, India

Geetha Palani Department of Physics, Faculty of Science and Technology, J.N.N. Institute of Engineering, Chennai, Tamil Nadu, India

N. Pavan Kumar Department of Sciences and Humanities, Matrusri Engineering College, Hyderabad, Telangana, India

R. Sakthivel Department of Mechanical Engineering, Amrita School of Engineering, Amrita Vishwa Vidyapeetham, Coimbatore, India

G.M. Shafiullah Discipline of Engineering and Energy, Murdoch University, Murdoch, Australia

Dipti Shukla O.P. Jindal University, Raigarh, Chhattisgarh, India

Vijay K. Sood University of Ontario, Institute of Technology, Oshawa, ON, Canada

Suresh Srinivasan Department of Electrical and Electronics Engineering, Annamacharya Institute of Technology and Science, Rajampet, Andhra Pradesh, India

D. Swathi Department of Physics, Telangana Social Welfare Residential Degree College for Women, Vikarabad, Telangana, India; Department of Physics, ISBM University, Nawapara (Kosmi), Gariyaband, Chhattisgarh, India

Musunuri Shanmukha Vardhan Department of Mechanical Engineering, Amrita School of Engineering, Amrita Vishwa Vidyapeetham, Coimbatore, India

Gerson O. Vargas-Godoy Universidad Tecnológica Centroamericana, Tegucigalpa, Honduras

Paul Vasquez-Miranda Escuela Politécnica Nacional, Universidad Politécnica Salesiana, Quito, Ecuador

Abhiram Vinod Department of Mechanical Engineering, Amrita School of Engineering, Amrita Vishwa Vidyapeetham, Coimbatore, India

Nand Kishore Yadav Department of Physics, ISBM University, Nawapara (Kosmi), Gariyaband, Chhattisgarh, India

An investigation of a photovoltaic system under distinguished thermal environment conditions

1

Anurag Koushal [1], *Yogesh K. Chauhan* [2], *Rupendra Kumar Pachauri* [3], *Thanikanti Sudhakar Babu* [4] *and Ali Q. Al-Shetwi* [5]
[1]Electrical Engineering Department, School of Engineering, Gautam Buddha University, Greater Noida, Uttar Pradesh, India; [2]Electrical Engineering Department, Kamla Nehru Institute of Technology, Sultanpur, Uttar Pradesh, India; [3]Electrical and Electronics Engineering Department, School of Engineering, University of Petroleum and Energy Studies, Dehradun, Uttarakhand, India; [4]Department of Electrical and Electronics Engineering, Chaitanya Bharathi Institute of Technology (CBIT), Hyderabad, Telangana, India; [5]Electrical Engineering Department, Fahad Bin Sultan University, Tabuk, Saudi Arabia

1. Introduction

Nowadays, humans are highly dependent on technologies which enhance the power demand from commercial and domestic applications. Due to limited storage capacity of fossil fuels, renewable energy (RE) sources such as solar energy, wind energy, biofuel, etc., have been explored and made significant contributions to meeting the world's growing energy demands. Furthermore, solar energy has no environmental impact and produces clean energy, which is the main reason for its popularity compared to other RE sources [1]. A normal photovoltaic (PV) module can utilize 6%−20% of solar radiation incident on it, dependent upon the type of solar cells used, and also turn depends upon the prevalent climatic conditions [2].

In Ref. [3], the authors discussed the working performance of hybrid photovoltaic/thermal (PV/T) systems in both energetic and exegetics modes. It is taken into consideration that the energy is always destroyed whenever there is a fluctuation in temperature. Thus, the increase in entropy is proportional to the destruction of energy that has occurred. With the observations made the authors were able to show an increase in electric power and solar cell conversion efficiency. They also listed various factors which could limit the efficient performance of the solar PV module, such as (a) reflection of light from the surface of the module, (b) fluctuation in solar irradiation, (c) series resistance in current flow, (d) shading effect due to surface top electric contacts, and (e) a lack of the optimal operating temperature. In Ref. [4], the authors enhanced the efficiency of the solar PV system experimentally by inculcating a water flow effect over the glass and water mass basins heat capacity. Numerical calculations revealed the enhanced efficiency of a solar still. In Ref. [5], the authors carried out numerical investigations for a heat transfer and energy conversion process. They were

Green Energy Systems. https://doi.org/10.1016/B978-0-323-95108-1.00001-X

able to show that, at low temperatures, there is an increase in thermal electrical generation for integrated PV systems per unit surface area. In Ref. [6], the authors calculated the effect of temperature variation and recorded the solar panel behavior. An analysis of solar module performance for temperature variation was made. The authors were able to show the improved performance of solar PV at low temperatures. The authors of Ref. [7] worked to improve the efficiency of solar PV systems using the floor heating system by passing a brine solution to cool the PV cells. This was done utilizing two different PV/T solutions for heat transfer analysis. In Ref. [8], the authors used halogen lamp radiation to calculate the effect of flow rate (volumetric) and the effect of angle of inclination on the performance of the solar PV system. In Ref. [9], the authors designed the PV/T system in such a way that the fluid flows on the top side of the PV panel in the fine gap between the solar cells of the PV panel. Using such a cooling method, the authors investigated the performance of the solar PV/T system. It was observed that there was an increment in the output power of the PV panel, leading to increased efficiency. In Refs. [10,11], the authors used two efficient strategies including a cooling methodology to increase the thermal electric performance of solar cells. Direct water cooling and cooling via a heated pipe were the methods used. This study was able to give the inverse relationship between PV cell temperature and terminal voltage. In Ref. [12], the authors conducted an experimental analysis of water flowing on top of the glass cover of a solar distillation system in a transient mode. The performance of the system was increased due to a significant difference between the temperatures of water and glass. In Ref. [13], the authors worked on the insulation effect on solar PV thermal collectors. The performance of a solar PV water collector in two modes, i.e., under partial and fully covered insulation, was discussed. An improvement in solar PV/T system efficiency involving impingement water jet cooling was discussed by the authors of Refs. [14,15]. Various modes such as radiation, electrical, optical, thermal, and jet cooling were included to understand the all-around PV system performance. The overall efficiency of the solar PV system was seen to obtain enhanced performance under the uniform cooling mode of working. In Ref. [16], a concentrated low-temperature fluid is routed to the rear of the panel through the heat exchanger. This heat was used for different domestic applications. Design parameters proposed in Ref. [17] were used for implementation. A review of PV/T hybrid solar technology is provided by the authors. Several theoretical and hardware models have been detailed. Data assessment and performance validation are carried out through implementation of beneficial technologies.

In Ref. [18], the authors performed an experimental analysis to show the effect of ambient temperature on PV system performance. The authors experimentally compared the correlation between temperature and efficiency for the correlation between wind speed and efficiency. In Ref. [19], the authors experimentally analyzed the seasonal change in the performance of a solar PV system. A comparative study of four seasons, namely monsoon, postmonsoon, winter, and summer, was carried out. The results showed that the lowest output was produced during the summer months (high ambient temperature) as compared to other months of operation. In Ref. [20], the authors provided a detailed review of the temperature dependency of PV efficiency along with its correlation and effect on the performance of a solar PV system throughout the world. In Ref. [21], the authors experimentally calculated the effect of dust on the performance of solar PV systems under partial shading during the operation. Experimentally derived results were compared to show the negative effect on solar PV efficiency.

The originality of this paper stems from the above-mentioned literature review, and aims to improve the performance of solar PV systems. Various experimental setups for thermal analysis are used to improve the working performance of both mono- and polycrystalline solar PV systems.

2. System description

A 50 W (polycrystalline) solar PV module is considered during this thermal experimental study. In this chapter, comprehensive analysis is carried out, such as (i) PV system performance at normal conditions, (ii) PV/T system performance with water flow, (iii) PV/T system performance with single-glazed water flow, (iv) PV/T system performance with reflector, (v) PV/T system performance with water flow and reflector, (vi) PV/T system performance with aerosol, (vii) PV/T system performance with combinations of aerosol and water flow, and (viii) PV/T system performance with aerosol, water flow, and reflector. The working performance is compared first within the system and then with the performance of the monocrystalline PV system. Fig. 1.1 depicts the experimental analysis schematic diagram. In this research, a comparative analysis of solar PV/T panels was undertaken, as well as other forms of thermal studies, in order to achieve optimum efficiency, voltage, current, and power of PV/T systems.

3. Modeling of PV system

For a PV cell, the total current I is calculated as the difference of photoelectric effect (I_L) and diode current (I_D), as expressed in Eq. (1.1). Moreover, open circuit voltage (V_{OC}) is expressed in Eq. (1.2) as follows:

$$I = I_l - I_D = I_l - I_0\left(e^{\frac{q(V+IR_s)}{AKT}}\right) \tag{1.1}$$

$$V_{OC} = \frac{AKT}{q}\ln\left(\frac{I_l}{I_0}\right) \tag{1.2}$$

where, I_O is diode saturation current (A) and q stands for electron charge (1.6×10^{-13} Coulombs). Moreover, k is the Boltzmann's constant (1.38×10^{-23} J/K) and T is used

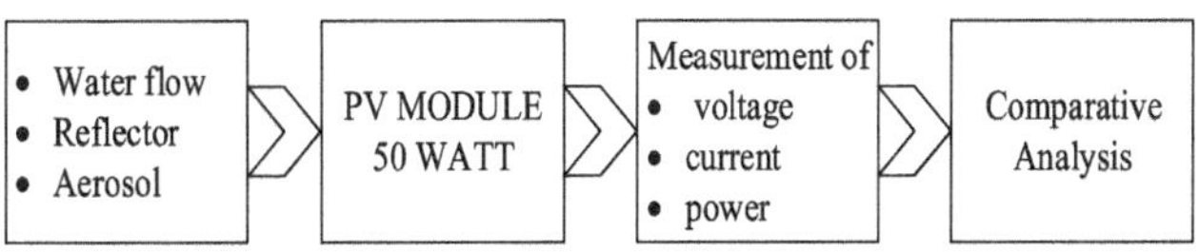

Figure 1.1 Schematic diagram of a PV/T system.

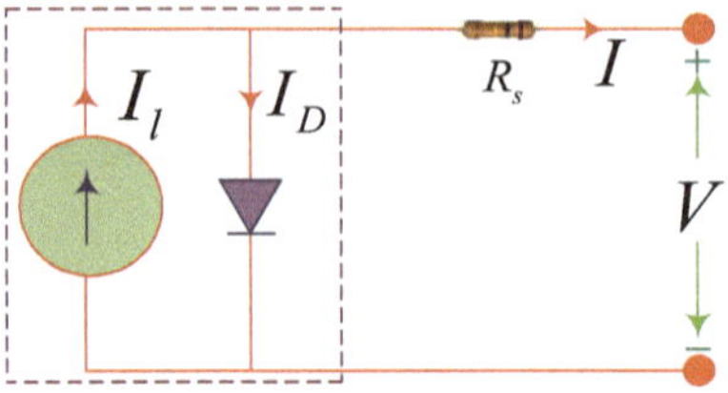

Figure 1.2 Simplified equivalent circuit model for a PV cell.

to show cell temperature (K). In addition, V shows the cell voltage (V), R_S represents the series resistance (Ω), and A represents the ideality factor. An electrical equivalent circuit is shown in Fig. 1.2.

4. Experimental analysis

A comprehensive study was performed with the established hardware arrangement under distinguished environmental conditions as follows.

4.1 Air-type solar PV/thermal system

This consists of air channels cut into the backside of the PV module, with air pushed to flow through the channels into the specified system, either naturally or violently. It is less efficient in locations where the temperature is higher than 20°C.

4.2 Water-type solar PV/thermal system

This is made up of channels cut into the backside of the PV modules, through which water is forced to flow into the system. To boost the rate of heat transmission, forced water circulation is used. These are more expensive than air-type systems but have a better output efficiency.

4.3 Refrigerant-type PV/thermal system

This increases the extraction of heat with the use of refrigerants. The rear of the panel is equipped with direct evaporation coils to evaporate coolant while the module flows. To improve the cooling rate of solar PV modules, reusable coolants of a high species heat capacity are utilized. Coolants circulate a lot, giving the system an additional advantage.

4.4 Bifluid-type PV/thermal system

Two working liquids are used by this system to keep the system cool. Water and air are both utilized this time. The heat is absorbed by two different pathways simultaneously. One fluid flows through the system; the other fluid is still moving. The heat-absorbing capacity of the solar PV module is simultaneously increased through the use of two fluids.

4.5 Glazed solar PV/thermal system

On the top of the solar PV modules is the entire system included in a collector with a glass coating. The thermal efficiency of a glass-covered system using a conventional solar PV/T system is increased.

Although thermal efficiency has improved, it has not increased electrical efficiency.

4.6 Unglazed solar PV/thermal system

The whole top of the PV module in this system is composed of glass. Compared to a normal solar PV/T system, this system delivers increased electricity efficiency without a glass cover. While electric efficiency is increased, there is a shortage of thermal efficiency.

In this chapter, working on glazed, unglazed, and water solar PV/T systems is performed on various experimental setups for both monocrystalline and polycrystalline PV panels. All the schematic diagrams and experimental setups for the analysis of the PV/T system are shown in Figs. 1.3A–C.

5. Results and discussion

5.1 PV system performance at normal conditions

The experimental system is arranged to observe the performance under normal conditions. The electrical responses in terms of PV system power, voltage, and current outputs are measured for a 2 -h study, which is summarized in Table 1.1.

Under the considered experimental conditions, a 50 W polycrystalline PV module with 156 × 156 mm size is considered. The temperature variation is taken from 31 to 44°C and the output voltage is dropped from 19.80 to 17.83 V, while the output power is dropped from 2.32650 to 2.03975 W. The performances in terms of voltage, power, and temperature of solar PV with respect to time are depicted in Figs. 1.4A and B.

5.2 PV/T system performance with only water flow

The experimental arrangement of the solar PV/T system is carried out with water flowing from the rear side of the panel. To maintain the flow of water properly, channels were made at the rear end of the panel, and water was made to flow through it. The electrical performance of the solar PV/T system is reported for 2 h at 10-min successive time intervals. The measurement data are depicted in Table 1.2.

Due to the temperature increase from 30 to 44°C, the voltage output dropped from 19.92 to 18.14 V, while the output power dropped from 2.34060 to 2.07522 W. The performance observation is carried out in terms of power, voltage, and temperature of the PV/T model with respect to time, and is shown in Figs. 1.5A and B.

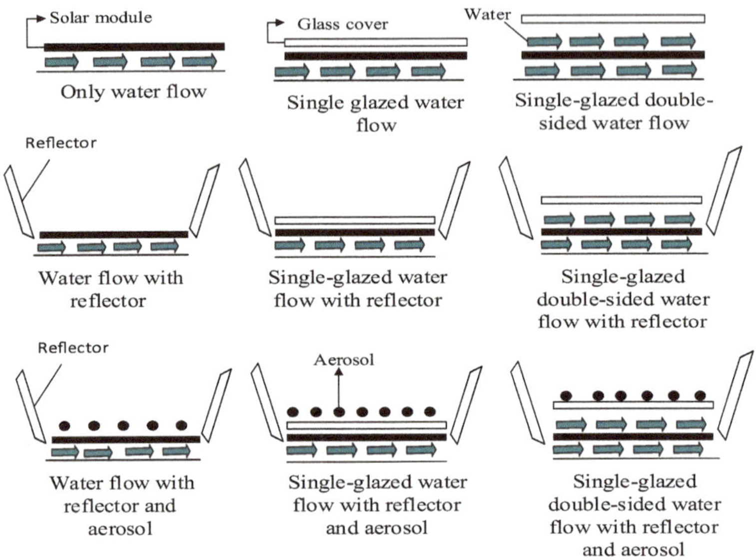

(A) Block diagrams of PV/T systems

(B) Set up of the polycrystalline solar PV panel

Figure 1.3 (A)–(C) Experimental setups for various types of thermal analysis.

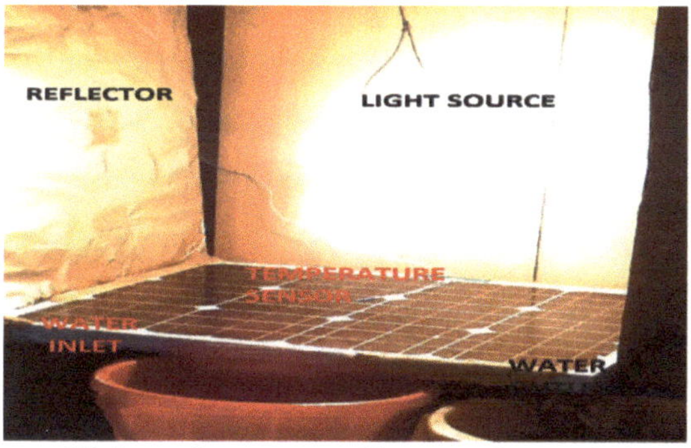

(C) Set up for monocrystalline water flow analysis

Figure 1.3 cont'd.

Table 1.1 PV system performance at ideal conditions.

Time (min)	Voltage (V)	Power (W)	Temp (°C)
0	19.80	2.32650	31
10	18.26	2.13642	35
20	17.98	2.09467	36
30	17.88	2.07944	36
40	17.87	2.06935	37
50	17.86	2.06462	37
60	17.86	2.05926	38
70	17.85	2.05275	39
80	17.85	2.04918	40
90	17.84	2.04625	41
100	17.84	2.04446	42
110	17.84	2.04268	43
120	17.83	2.03975	44

5.3 PV/T system performance with single-glazed water flow

The PV/T system is experimentally designed to function with water coming from the back of the panel, while the upper part of the panel was isolated by one clear glass. The thickness of the glass is 2 mm. Channels are created at the back of the panel to ensure the correct water flow and water is drained from the panel. For 2 h, at 10–min intervals, are the voltage, power, and temperature of the PV/T system are measured, and measurement data are reported in Table 1.3.

The output voltage reduced from 19.47 to 18.07 V with an increase in temperature from 31 to 48°C. The output power fell to 2.085278 from 2.307195 W. The performances of voltage, power, and temperature of solar PV/T with respect to time are shown in Figs. 1.6A and B.

Figure 1.4 (A) and (B) PV performance at normal conditions.

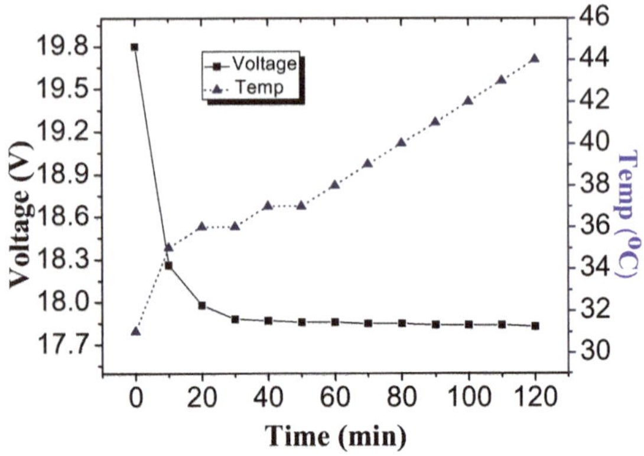

(A) Voltage and temperature characteristics

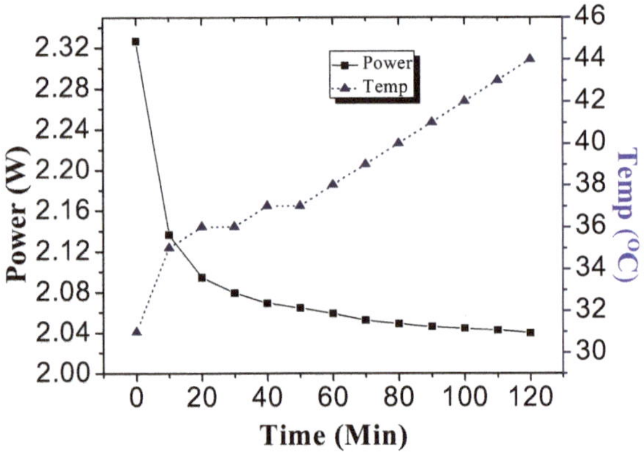

(B) Power and temperature characteristic

Table 1.2 Electrical performance PV/T system with only water flow.

Time (min)	Voltage (V)	Power (W)	Temp (°C)
0	19.92	2.34060	30
10	18.97	2.21949	33
20	18.52	2.15758	34
30	18.38	2.13759	35
40	18.24	2.11219	35
50	18.21	2.10508	37
60	18.19	2.09731	38
70	18.19	2.09185	39
80	18.19	2.08821	40
90	18.16	2.08295	42
100	18.14	2.07884	43
110	18.14	2.07703	44
120	18.14	2.07522	44

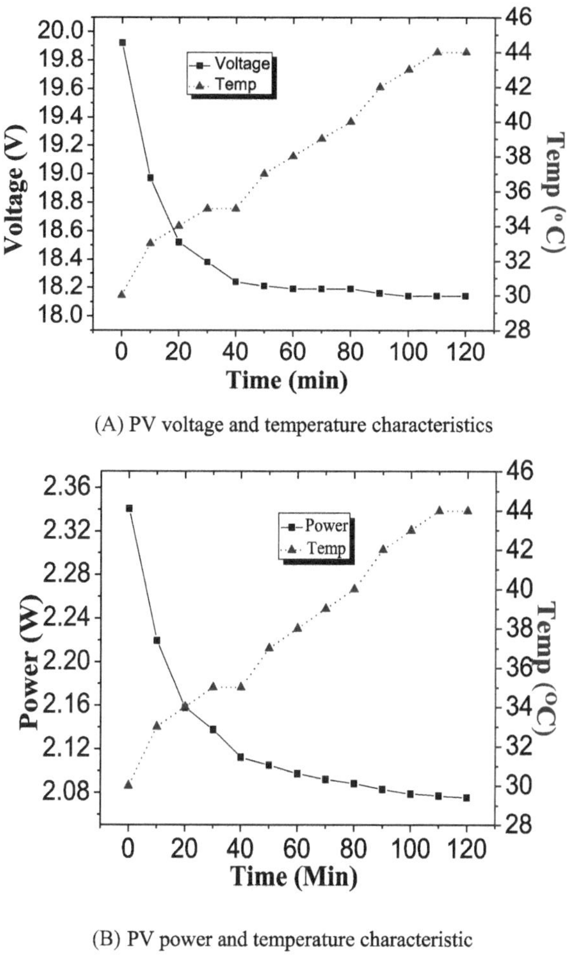

Figure 1.5 (A) and (B) PV/T system performance with only water flow.

Table 1.3 PV/T system performance with single-glazed water flow.

Time (min)	Voltage (V)	Power (W)	Temp (°C)
0	19.47	2.307195	31
10	18.42	2.173560	36
20	18.27	2.146725	38
30	18.13	2.126649	40
40	18.10	2.114080	43
50	18.09	2.109294	45
60	18.09	2.103867	46
70	18.08	2.097280	46
80	18.08	2.093664	47
90	18.06	2.089542	47
100	18.06	2.087736	48
110	18.07	2.087085	48
120	18.07	2.085278	48

Figure 1.6 (A) and (B) PV/T system performance with single-glazed water flow analysis.

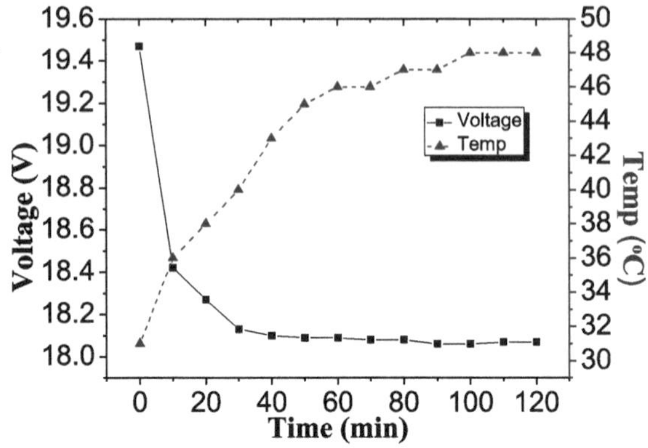

(A) PV voltage and temperature characteristic

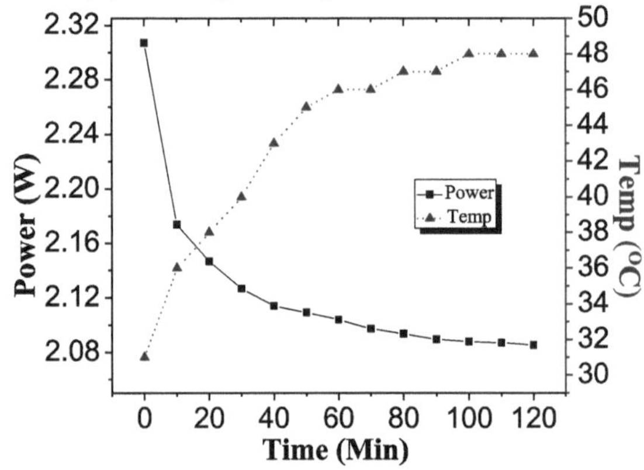

(B) PV power and temperature characteristic

5.4 PV/T system performance with reflector

The experimental design of the PV/T model is organized to work perpendicular to the surface of the panel with a reflector set (at optimal height). The readings of voltage, power, and temperature of solar PV/T system are measured for 2 h at 10–min successive intervals, and the recorded data are provided in Table 1.4.

With the increment in temperature from 31 to 48°C, the output voltage is reduced from 20.39 to 17.90 V. In addition, the output power is decreased from 2.50797 to 2.07282 W. The performances of voltage, power, and temperature of the solar PV/T with respect to time are shown in Figs. 1.7A and B.

5.5 PV/T system performance with water flow and reflector

The experimental setup of the PV/T system is arranged for water to flow along with the reflectors from the back of the panel (on optimum height). Channels were created at the back of the panel to ensure the correct water flow and water was drained from the panel. For 2 hours, at 10 min repeated time intervals, solar PV/T system voltage, power, and temperature are measured and the measured data are reported in Table 1.5.

Table 1.4 Performance of PV/T system with reflector.

Time (min)	Voltage (V)	Power (W)	Temp (°C)
0	20.39	2.50797	31
10	19.27	2.35094	37
20	18.31	2.21551	42
30	18.12	2.17440	42
40	18.01	2.13419	45
50	17.95	2.11810	45
60	17.92	2.10560	45
70	17.92	2.10202	47
80	17.91	2.09189	47
90	17.91	2.08831	47
100	17.91	2.08293	48
110	17.91	2.07756	48
120	17.90	2.07282	48

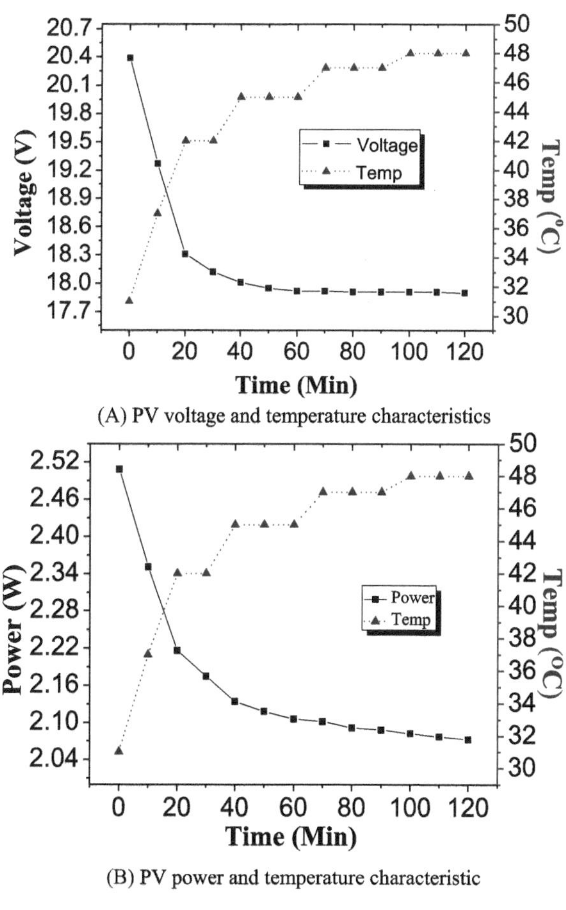

Figure 1.7 (A) and (B) PV system performance with reflector.

Table 1.5 PV/T system performance with water flow and reflector.

Time (min)	Voltage (V)	Power (W)	Temp (°C)
0	20.60	2.863400	29
10	19.11	2.853100	38
20	18.88	2.637180	42
30	18.63	2.596000	45
40	18.51	2.557899	45
50	18.47	2.532168	46
60	18.44	2.523002	47
70	18.41	2.513372	47
80	18.40	2.503760	47
90	18.37	2.498720	48
100	18.37	2.492809	48
110	18.36	2.490972	48
120	18.35	2.487780	48

The output voltage fell from 20.60 to 18.35 V with a temperature rise from 29 to 48°C. The output power fell to 2.487780 from 2.863400 W. In Figs. 1.8A and B, the performances of solar PV/T voltage, power, and temperature are illustrated.

5.6 PV/T system performance with aerosol condition

Aerosol dusting has been developed for the experimental layout of the PV solar system. The influence of dust and the heat effect are examined in this system. To do this, a sieve was used to scatter 10 g of construction sand on the top of the panel equally. Measurements of PV/T solar system voltage, power, and temperature are taken over a period of 2 h at 10–min intervals, as shown in Table 1.6.

The output voltage decreased from 17.85 to 17.52 V with an increase in temperature from 32 to 44°C. The power fell to 2.004288 from 2.075955 W. In Figs. 1.9A and B, the performances in terms of solar PV/T voltage, power, and temperature with regard to time are illustrated.

5.7 PV/T system performance with aerosol and water flow

The experimental arrangement of the solar photovoltaic systems is designed to work along with rear water supplies with aerosol dusting. The effect of dust and the thermal effect are examined in this system. In order to do this, a sieve was applied to 10 g of construction sand on the top side of the panel, while the rear of the panel was supplied by a water channel. For 2 hours at 10 min intervals the voltage, energy, and temperature of the PV/T solar system were measured, and the measured data are provided in Table 1.7.

The output voltage dropped from 17.97 to 17.63 V with an increase of temperature from 31 to 44°C. The output power fell to 2.0327 from 2.1204 W. The voltage, power, and temperature performances of the solar PV/T are shown in relation to time in Figs. 1.10A and B.

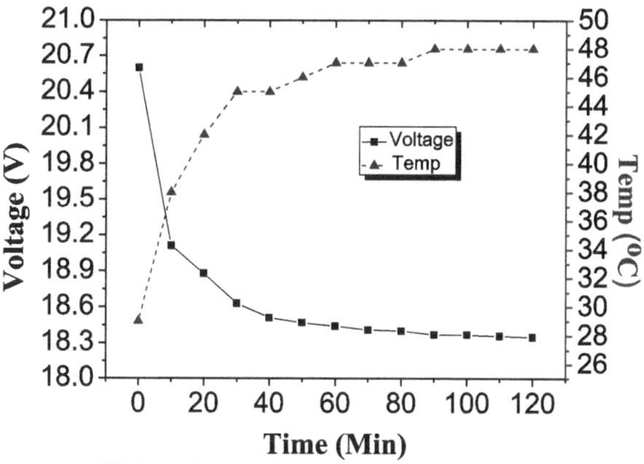

(A) PV voltage and temperature characteristics

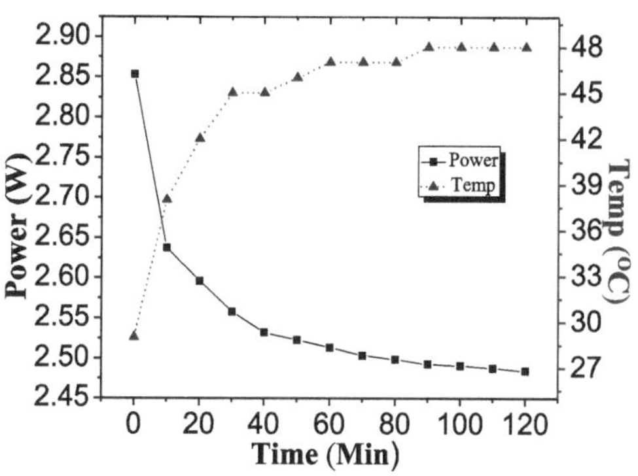

(B) PV power and temperature characteristic

Figure 1.8 (A) and (B) PV/T system performance with water flow with reflector.

Table 1.6 PV/T system performance with aerosol condition.

Time (min)	Voltage (V)	Power (W)	Temp (°C)
0	17.85	2.075955	32
10	17.62	2.035110	34
20	17.59	2.029886	35
30	17.58	2.026974	36
40	17.57	2.024064	37
50	17.56	2.021156	37
60	17.56	2.019400	38
70	17.56	2.017644	38
80	17.55	2.014740	39
90	17.55	2.012985	41
100	17.54	2.010084	41
110	17.53	2.007185	43
120	17.52	2.004288	44

Figure 1.9 (A) and (B) PV/T system performance with aerosol dusting.

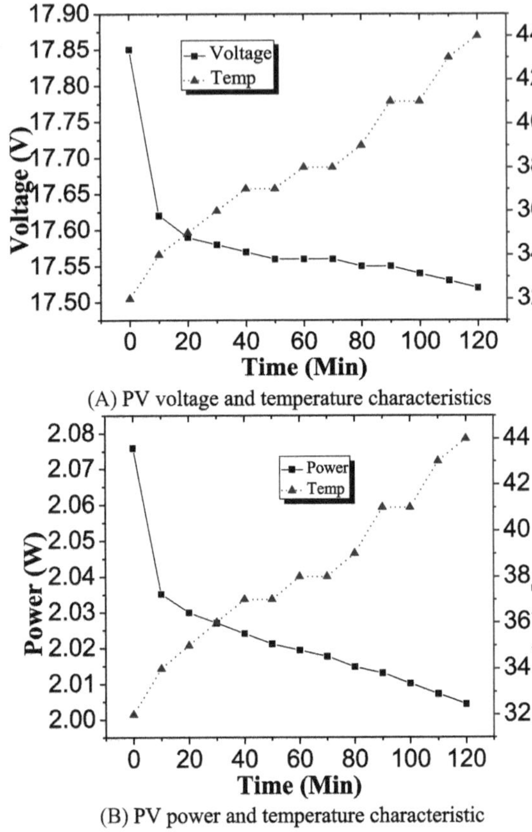

Table 1.7 PV/T system performance with aerosol and water flow.

Time (min)	Voltage (V)	Power (W)	Temp (°C)
0	17.97	2.120460	31
10	17.76	2.086800	32
20	17.71	2.077383	33
30	17.69	2.066192	34
40	17.68	2.061488	35
50	17.68	2.056184	36
60	17.67	2.049720	37
70	17.66	2.045028	39
80	17.66	2.043262	41
90	17.66	2.041496	42
100	17.64	2.037420	43
110	17.64	2.035656	44
120	17.63	2.032739	44

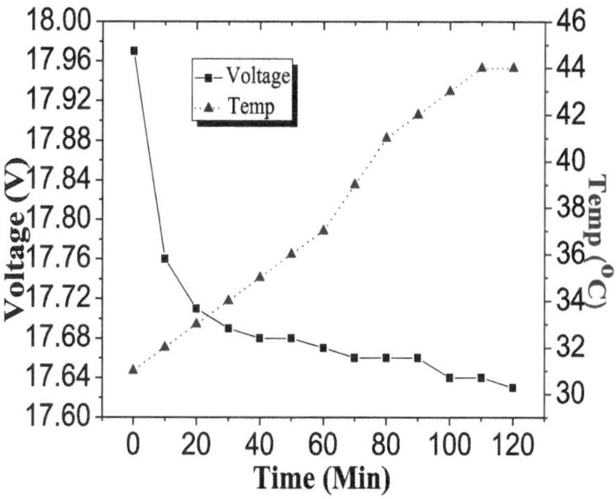

(A) PV voltage and temperature characteristics

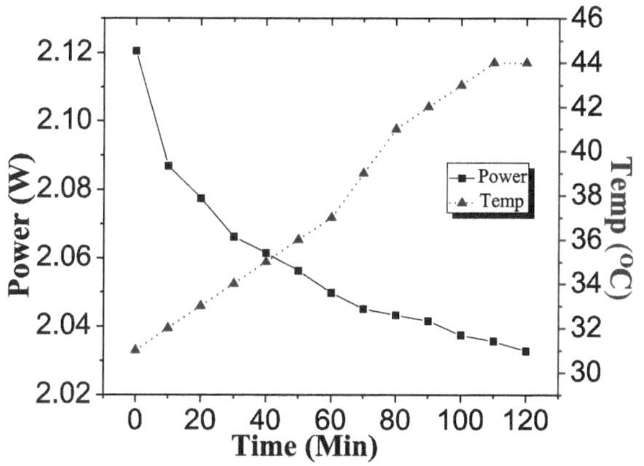

(B) PV power and temperature characteristic

Figure 1.10 (A) and (B) PV/T system with aerosol and water flow.

5.8 PV/T system performance with aerosol and water flow with reflector

In addition to the rear side water, the experimental arrangement of the solar PV/T system is arranged to work in perpendicular configuration with aerosol pollution. The effect of dust and the thermal effect are examined in this system. In order to do this, a sieve was applied to 10 g of construction sand on the top side of the panel, while the rear of the panel was supplied by a water channel. For 2 hours, at 10–min intervals voltage, power, and temperature of the solar PV/T system were measured, and the measured data are provided in Table 1.8.

The output voltage dropped from 18.01 to 17.73 V with an increase in temperature of between 30 and 48°C. The output power decreased to 2.028312 from 2.116175 W. The voltage, current, and energy performances of the solar energy PV/T are shown in relation to time in Figs. 1.11A and B.

Table 1.8 Performance of PV system with aerosol and water flow with reflector.

Time (min)	Voltage (V)	Power (W)	Temp (°C)
0	18.01	2.116175	30
10	17.91	2.095470	36
20	17.82	2.076030	43
30	17.80	2.070140	44
40	17.79	2.060082	45
50	17.77	2.054212	46
60	17.77	2.048881	46
70	17.76	2.042400	47
80	17.76	2.038848	47
90	17.74	2.034778	47
100	17.74	2.033004	48
110	17.73	2.030085	48
120	17.73	2.028312	48

Figure 1.11 (A) and (B) PV/T performance with aerosol and water flow with reflector.

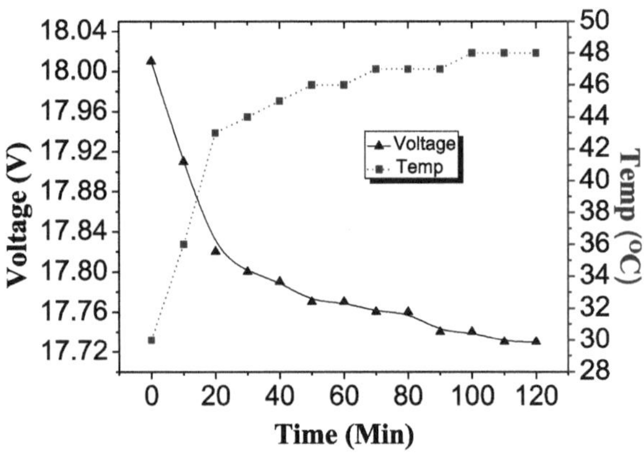

(A) PV voltage and temperature characteristics

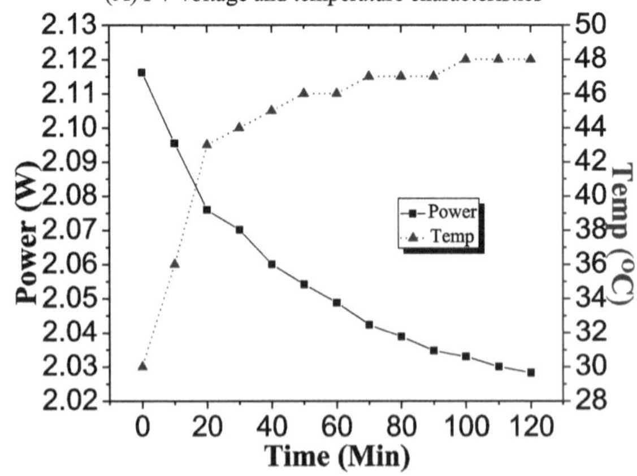

(B) PV power and temperature characteristic

5.9 Polycrystalline comparison between all thermal conditions for PV/T system

The comparison graphs of polycrystalline PV/T Figs. 1.12A–G and the performance comparisons of all the PV/T are tabulated in Table 1.10.

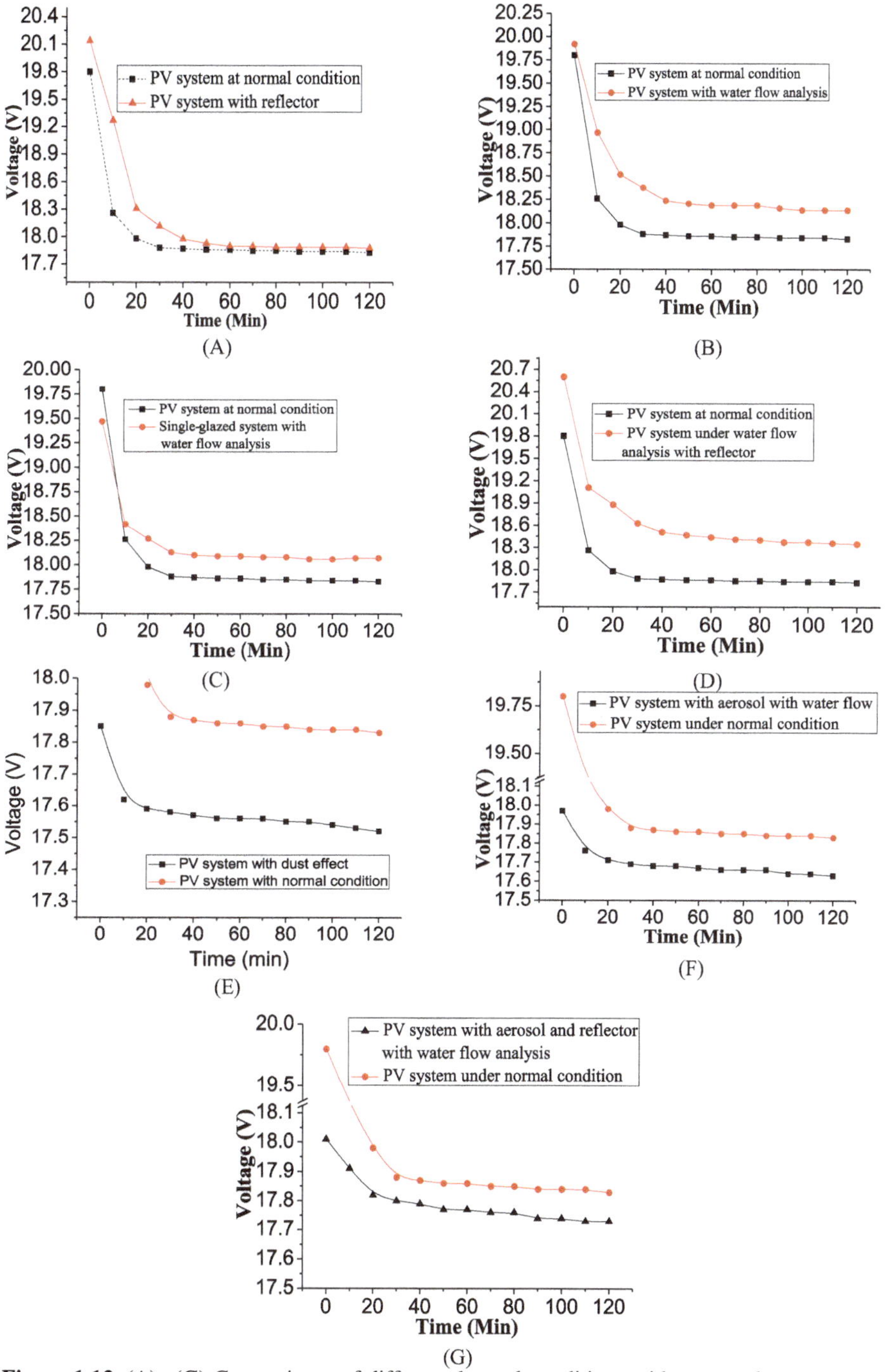

Figure 1.12 (A)–(G) Comparisons of different thermal conditions with a normal polycrystalline PV system.

5.10 *Performance comparison between polycrystalline and monocrystalline PV/T systems*

The experimental results obtained for the monocrystalline solar PV system in Ref. [1] are compared with the experimental results obtained for the polycrystalline solar PV system for different experimental setups, as shown in Figs. 1.13A–E. The effect of

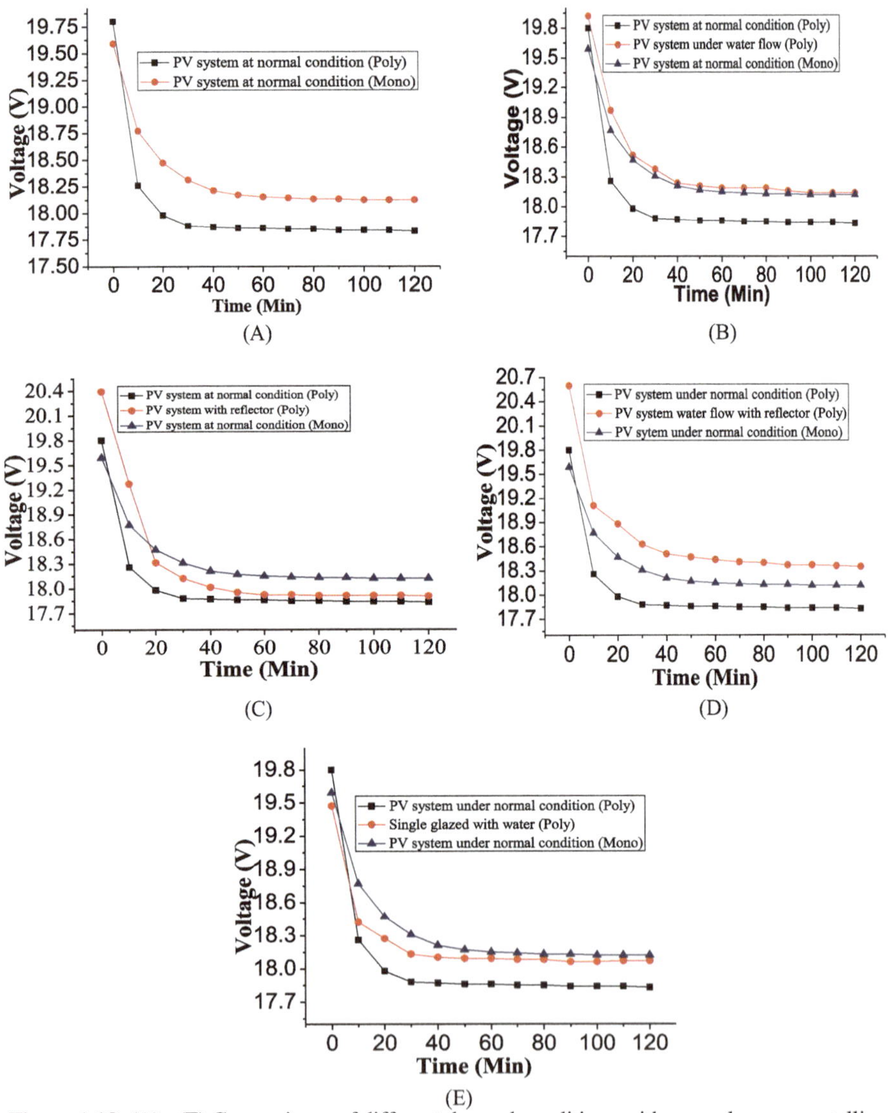

Figure 1.13 (A)–(E) Comparisons of different thermal conditions with normal monocrystalline and polycrystalline PV systems.

the reflector, water flow, single–glazed system with water flow, water flow with reflector and dust effect, dust with water flow, and dust and reflector with water flow are compared with the PV system at normal condition.

The tabulated stability performance of different solar PV/T systems is shown in Table 1.9.

Table 1.9 Performance comparison of PV/T systems.

Performance of PV/T systems	**Voltage at $T = 0$ min**	**Change in voltage ($T = 0$)**	**Voltage at $T = 120$ min**	**Change in voltage ($T = 120$)**	**Stability after 120 min**
PV system with only water flow	19.92	+0.6061	18.14	+1.7386	Positive
PV system with single–glazed water	19.47	−1.6667	18.07	+1.3460	Positive
PV system with reflector	20.39	+2.9798	17.90	+0.3925	Positive
PV system with water flow and reflector	20.60	+4.0404	18.35	+2.9164	Positive
PV system with aerosol dusting	17.85	−9.8485	17.52	−1.7386	Negative
PV system with aerosol and water flow	17.97	−9.2424	17.63	−1.1217	Negative
PV system with aerosol and water flow with reflector	18.01	−9.0404	17.73	−0.5608	Negative

Table 1.10 Comparative performance of all thermal environmental conditions.

Time (min)	$\Delta T = (T_0 - T_{120})$	Voltage (V)	% Change in V	Power (W)	% Change in W	Temp (° C)	Δ°C = (°C_0 − °C_{120})
Performance of PV/T system at normal conditions							
0	120	19.80	9.9495	2.32650	12.3254	31	13
120		17.83		2.03975		44	
Performance of PV/T system with only water flow							
0	120	19.92	8.9357	2.34060	11.3381	30	14
120		18.14		2.07522		44	
Performance of PV/T system with single−glazed water flow							
0	120	19.47	7.0281	2.307195	9.6185	31	17
120		18.07		2.085278		48	
Performance of PV/T system with reflector							
0	120	20.39	12.2119	2.50797	17.3507	31	17
120		17.90		2.07282		48	
Performance of PV/T system with water flow and reflector							
0	120	20.60	10.9223	2.86340	13.1179	29	19
120		18.35		2.48778		48	
Performance of PV/T system with aerosol dusting							
0	120	17.85	1.8487	2.075955	3.4522	32	12
120		17.52		2.004288		44	
Performance of PV/T system with aerosol and water flow							
0	120	17.97	1.8920	2.120460	4.1369	31	13
120		17.63		2.032739		44	
Performance of PV/T system with aerosol and water flow with reflector							
0	120	18.01	1.5546	2.116175	4.1519	30	18
120		17.73		2.028312		48	

6. Conclusion

In this chapter, the solar PV/thermal performance of polycrystalline panels has been investigated comprehensively under novel conditions for combinations of water flow, reflectors, and aerosols. The following are the findings of this study:

- On experimentation, it was found that the solar PV/T system with water flow with a reflector was able to provide maximum positive stability of +2.9164% followed by the solar PV system with only water flow analysis +1.7386%, solar PV/T system with the single–glazed system, and solar PV/T system with reflector, in that order.
- For the aerosol system, it was found that the solar PV/T system with aerosol dusting provided maximum negative stability, but on the application of different solar PV/T systems such as a solar PV/T system with aerosol with water flow and solar PV/T system with aerosol and water flow with reflector it tended to show a decrease in its negative stability, in that order, respectively.
- In comparison with the performance of similar systems of monocrystalline solar panels, it was observed that though normally the performance of polycrystalline systems is lower than that of monocrystalline, on the application of different PV/T systems, the performance of a polycrystalline system tends to approach the performance of a monocrystalline system (at normal condition).

Appendix

See Table 1A.1.

Table 1A.1 Datasheet of commercial PV model 50 W manufactured by Indo Solar.

Power (W)	**50 (polycrystalline)**
Series resistance (Ω)	1.138
Parallel resistance (Ω)	56.789
Cell number	36
Ambient temp (°C)	25.0
I_{mp} (A)	2.988
V_{mp} (V)	18.34
FF	0.751
P_{mp} (W)	54.8083
I_{sc} (A)	3.277
V_{oc} (V)	22.26
Efficiency	18.77
U (v)	0.00
Irradiance (W/m^2)	1000.0

References

[1] P.R. Satpathy, S. Babu Thanikanti, A.H.O.U.S.E.I.N. Mahmoud, R. Sharma, B. Nastasi, A TCT-SC hybridized voltage equalizer for partial shading mitigation in PV arrays, IEEE Transactions on Sustainable Energy 12 (4) (2021) 1261–1281.

[2] A. Gupta, P. Kumar, R.K. Pachauri, Y.K. Chauhan, Effect of environmental conditions on single and double diode PV system: a comparative study, International Journal of Renewable Energy Resources 4 (4) (Dec. 2014) 849–858.

[3] E. Radziemska, Performance Analysis of a Photovoltaic-Thermal Integrated System vol 2009, Hindawi Publishing Corporation, 2009, pp. 1–6.

[4] S.A. Lawrence, et al., Effect of heat capacity on the performance of solar still with water flow over the glass cover, Energy Conversion and Management 30 (3) (1990) 277–285.

[5] L. Jing, G. Pei, Y. Li, J. Ji, Design and performance analysis of low temperature solar thermal electric generation integrated PV cell, in: Proceedings IEEE Conference on Asia-Pacific and Power and Energy Engineering at Chengdu on 28–31 March, 2010, pp. 1–4.

[6] A.L. Malta, et al., Development of a thermal visor to analyze the influence on temperature in the efficiency of solar panel, in: Proceedings IEEE Conference on Innovative Smart Grid Technologies Latin America, at Sao Paula on 15–17 April, 2013, pp. 1–5.

[7] C.S. Polo Lopez, et al., Testing of a cost-effective photovoltaic thermal hybrid solar collector prototype, in: Proceedings IEEE Conference on Photovoltaic Specialist at Austin, on 3–8 June, 2012, pp. 479–484.

[8] R. Ihaddadene, et al., The effects of volumetric flow rate and inclination angle on the performance of a solar thermal collector, Energy Conversion and Management 78 (2014) 931–937.

[9] S. Dubey, et al., Performance analysis of innovative top cooling thermal photovoltaic (TPV) modules under tropics, in: Proceedings International Conference on Energy Sustainability, June 26–30, 2016, pp. 1–5.

[10] M.E. Beck, S. Wiedeman, R. Huntington, J. VanAlsburg, E. Kanto, R. Butcher, J.S. Britt, Advancements in flexible CLGS module manufacturing, in: Proceedings IEEE International Conference on Photovoltaic Specialist at USA on 3–7 January, 2007, pp. 211–214.

[11] M.A. Farhat, Improvement the thermal electric performance of photovoltaic cells by cooling and concentration techniques, in: Proceedings IEEE International Conference on Power Engineering at Bristol, on 8 September, 2004, pp. 623–628.

[12] V. Bapeshwar, G.N. Tiwari, Effect of water flow over the glass on the performance of a solar still coupled with a flat plate solar collector, International Journal of Solar Energy 2 (1984) 277–288.

[13] M.F. Mohammed, N.A. Rahim, M. Hasanuzzaman, A. Rivai, Effect on insulation of photovoltaic thermal water collector, in: Proceedings IEEE Conference on Clean Energy and Technology at Lankgkawi, on 18–20 November, 2013, pp. 307–311.

[14] M.S. Haitham, A.A. Bahaidarah, B. Baloch, P. Gandhidasan, Modeling and comparative analysis of jet impingement cooling and conventional channel cooling for photovoltaic strings, in: Proceedings IEEE Conference on Photovoltaic Specialist Conference at Denver, on 8–13 June, 2014, pp. 748–753.

[15] H.M.S. Bahaidarah, Experimental performance investigation of uniform and non-uniform cooling techniques for photovoltaic systems, in: Proceedings IEEE Conference on Photovoltaic Specialist Conference at New Orleans, on 14–19 June, 2015, pp. 1–4.

[16] R. K Koech, H.O. Ondieki, J.K. Tonwi, S.K. Rotich, A steady state thermal model for photovoltaic/thermal system under various condition, International Journal of Scientific and Technology Research 1 (11) (December 2012) 1–5.
[17] T.T. Chow, A review on photovoltaic/thermal hybrid solar technology, Applied Energy 87 (2010) 365–379.
[18] T. Bhattacharya, et al., Effects of Ambient Temperature and Wind Speed on Performance of Monocrystalline Solar Photovoltaic Module in Tripura, India, Hindawi Publishing Corporation, 2014.
[19] M.S. Vasisht, et al., Performance of Solar Photovoltaic Installations: Effect of Seasonal Variations, Solar Energy, 2016, pp. 39–46.
[20] S. Dubey, et al., Temperature dependent photovoltaic (PV) efficiency and its effect on PV production in the world: a review, Energy Procedia 33 (2013) 311–321.
[21] P. Yadav, et al., Investigations on the effects of partial shading and dust accumulation on PV module performance, in: Proceedings of International Conference on Intelligent Communication, Control and Devices, 2016, pp. 1005–1012.

Design and simulation of a renewable-based sustainable electrification system for a water purification plant

2

Murugaperumal Krishnamoorthy [1], *Karuppiah Natarajan* [1], *Praveen Kumar Balachandran* [1] *and Suresh Srinivasan* [2]
[1]Department of Electrical and Electronics Engineering, Vardhaman College of Engineering, Hyderabad, Telangana, India; [2]Department of Electrical and Electronics Engineering, Annamacharya Institute of Technology and Science, Rajampet, Andhra Pradesh, India

1. Introduction

Energy generation from fossil fuels leads to greenhouse gas emissions, which increases global warming [1–3]. The integration of renewable energy with existing energy utilities could mitigate the causes of climatic change resulting in air pollution reduction, better human health, low-cost energy generation, and an energy balance between rural and urban areas. This provides a solution for power scarcity and aids rural sustainability [4–6].

Owolabi et al. [7] examined the energy efficiency of residential buildings in terms of proper energy efficiency measures. The existing building is retrofitted by advanced building technology and a regularity strategy and it was found that typical residential buildings could save nearly 54 GJ of fuel every year. The forecast mitigation emission of GHG is 2.71 tonnes of CO_2. Owolabi et al. [8] attempted to derive the statement of environmental stability in terms of solar resources involvement with Nigerian zoned climatic dates. A level of 5.96 kWh/m^2/d of solar irradiation could produce 11,385 MWh of energy for the zoned grid for environment sustainability. The existing gas-fired power plants usage has to be reduced using the proposed worthy sustainable energy generation technology.

Moya et al. [9] have developed a feasibility report for a geothermal power plant and its positive impacts on society over fossil fuel-fired energy generation. A typical 22 MW geothermal plant costs 114.3 million USD and has the capability to mitigate 184,000 tonnes of CO_2 footprint over its life span. Lee et al. [10] submitted the research outcomes of a combined hybrid renewable heating system having more technoeconomical advancements over standalone renewable energy generating systems. They considered many indexing parameters related to optimal design of hybrid energy systems which include different levels of renewable resources sensitivity. Wieda et al. [11] examined the grid-connected solar plant performance for optimal peak load management purposes. They considered roof and ground level-mounted PV systems with different seasonal variations for attaining a high level of predilection accuracy.

Green Energy Systems. https://doi.org/10.1016/B978-0-323-95108-1.00002-1

Ramli et al. [12] published a comparative study report of the top-rated renewable energy simulation tools such as HOMER and Retscreen. Each tool has its pros and cons based on the usage of system constraints and the level of component integration. Aristizábal et al. [13] used the Retscreen software for power quality analysis of a building-integrated PV system and its evaluation. IEEE 929-2000 standard guidelines were followed to estimate the optimal configuration of components for a lower life cycle cost of the system greater GHG pollutant mitigation. Dikgale et al. [14] developed a water purification system with the help of direct heat from solar energy. The flat plate collectors and water circulation system have been combined for a water purification system. Nur et al. [15] designed and constructed a solar water purifier with the help of a filtering and boiling mechanism to remove dirt from the stored water. Arsanto et al. [16] proposed a water treatment plant powered by a solar resource for remote area applications by Indonesian communities. The economic performance and rate of investment returns were derived and compared against the current system. Prakash et al. [17] developed a microcontroller-based water purification system powered by renewable energy. The demand and supply relationship was traced and water purification processes were carried out.

From the field state of the art, the novelty of this work can be defined as (a) what level of impacts could be reacting on the existing system, when incorporating a renewable source retrofit, (b) the level of system sensitivity and designing a range of optimal energy configuration of electricity utilities, and (c) the course of work covers both the areas of energy demand and resource availability, including an understanding of comparative cost savings among different operating conditions. In this part of study, Section 2. defines the system modeling and problem statement. Then, Section 2. deals with the proposed system architecture which includes the chosen research methodology and merit claiming system constraints. In Section 2., the results and a discussion of the predicted outcomes are explained. The final section provides information about the conclusion of the case study and future scope of this work.

2. Case study modeling and problem statement

The main aim of this case study is to develop a solar PV-powered water purification plant for sustainability analysis. The existing system takes power from conventional energy sources. To mitigate the adverse consequences of using fossil fuels, the solar energy retrofit proposal was raised and its overall performance verified by standard sustainability indexes.

2.1 System problem statement

The available campus water purification plant is operated by an electric motor pump set with reverse osmosis purification components. During plant operation, a 1.5 HP motor pump set is operated for 10 h daily. The total water-storing capacity before the purification process is 20,000 L and, after the purification process, it is

15,000 L. The total energy needed per day is approximately 18.65 kWh. The monthly energy consumption is 559.5 kWh, and annually this is 6807.25 kWh. The purchasing cost of one unit under a commercial tariff is INR 7 per unit. In the case of the proposed solar water-purifying system, 1.5 kW of solar panels is used with a 40 Ah, 12 V battery type of energy storage system. The typical RO purification process has to be followed.

2.2 Plant energy modeling

The total energy required to operate the RO plant has been expressed in Eq. (2.1):

$$Q_P = (V_p)(\rho_p)(C_{c,p})(\Delta T_P) \tag{2.1}$$

where V_p is the volume of process fuel in m^3, the density of fluid is represented by ρ_p in kg/m^3, the specific heat of fluid is C_p in kJ/kG/k, and the process increasing temperature is represented as ΔP in °C. The time needed for the temperature rise for the purification process is expressed by Eq. (2.2):

$$T(t) = \frac{Q_p * 1000}{\text{power of heating coil} * 60} \tag{2.2}$$

where, Q_p is the energy requirement for the purification process. The solar PV power generation can be expressed by Eq. (2.3):

$$P_{spv}(t) = Y_{spv} \times D_{spv}\left(\frac{R_t}{R_{tSTC}}\right)\left[1 + \partial_p(T_c - T_{cSTC})\right] \tag{2.3}$$

where Y_{Spv} is the solar PV modulerated capacity, D_{spv} is the panel derating factor, R_t is the solar irradiation, T_c is the cell temperature, and R_{tSTC} and T_{cSTC} are the standard test conditions for solar PV, respectively, whereas battery charging and discharging are expressed as [$(\Delta P(t) > 0)$ and $(\Delta P(t) < 0)$] which is generalized by Eq. (2.4):

$$P_{bt}(t+1) = P_{bt}(t) + \left(\frac{\Delta P(t)}{V_{dc}}\right)\eta_{rt}\Delta t \tag{2.4}$$

The total energy for the chosen system from the integrated hybrid system is given by Eq. (2.5):

$$P_{tot}(t) = \sum_{n=1}^{sn} P_{pv}(t) + \sum_{w=1}^{sm} P_{Gw}(t) + \sum_{btt=1}^{so} P_{btt}(t) \tag{2.5}$$

where (n, m, o) is the matrix of optimization search space of the energy generator components. P_{pv} is solar power, P_{Gw} is grid power, and P_{btt} is the battery backup power.

2.3 Renewable resource analysis

The chosen case study area was in South India, in the central part of Telangana state. It is located between a latitude of 17.5 N and longitude of 78.5 E; it is an extremely hot–humid type of climatic zone. The available average solar radiation is 5 kWh/m^2/d. The existing annual average temperature range is between 21.6 and 33°C. Fig. 2.1 and Table 2.1 illustrate the same.

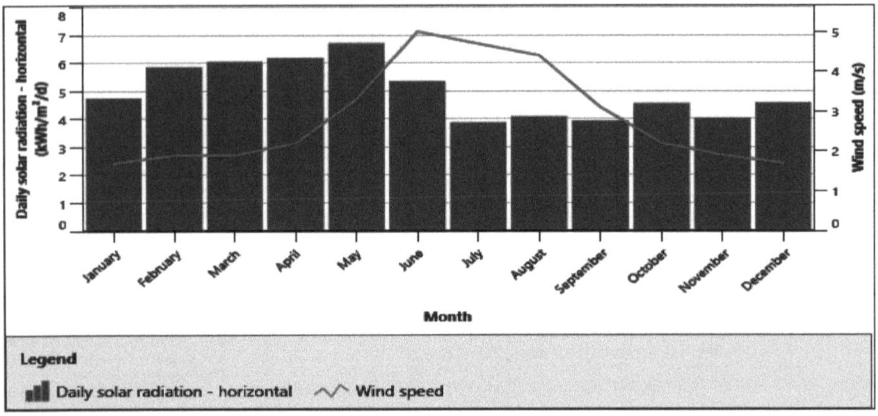

Figure 2.1 Renewable resource availability in the proposed site location.

Table 2.1 Monthly resource availability on the selected site.

	Temperature (°C)	% Humidity	Pressure (kPa)	Wind speed (m/s)
January	22.2	59.8	95.0	1.7
February	25.1	51.0	94.7	1.9
March	28.4	44.7	94.5	1.9
April	31.5	43.8	94.2	2.2
May	33.0	42.4	93.9	3.3
June	29.3	62.9	93.8	5.0
July	27.0	74.4	93.9	4.7
August	26.2	78.4	94.0	4.4
September	26.6	75.6	94.2	3.1
October	25.7	71.2	94.5	2.2
November	23.2	64.3	94.8	1.9
December	21.6	60.7	95.0	1.7
Annual	**26.7**	**60.8**	**94.4**	**2.8**

3. Research methodology—Retscreen simulation

The feasible configuration of the proposed hybrid renewable energy system is constructed and its performance verified using Retscreen expert tools. Here, four stages of processing steps are followed, including virtual energy analysis, smart project identifier, financial risk assessor, and performance tracker. The load assessment, technological availability, and component assessment are considered in the first two stages. The result feasibility and decision assessments are the main key indexes of the next level of simulation. The detailed simulation process is illustrated in Fig. 2.2.

Fig. 2.3 illustrates the hybrid energy retrofit model of a solar water-purifier plant. The water-purifying process is carried out through available solar, grid, and battery powers. The performance improvement benchmarks are fixed by the Retscreen expert model. From the simulation outcomes, the system feasibility and the energy management report are developed and analyzed.

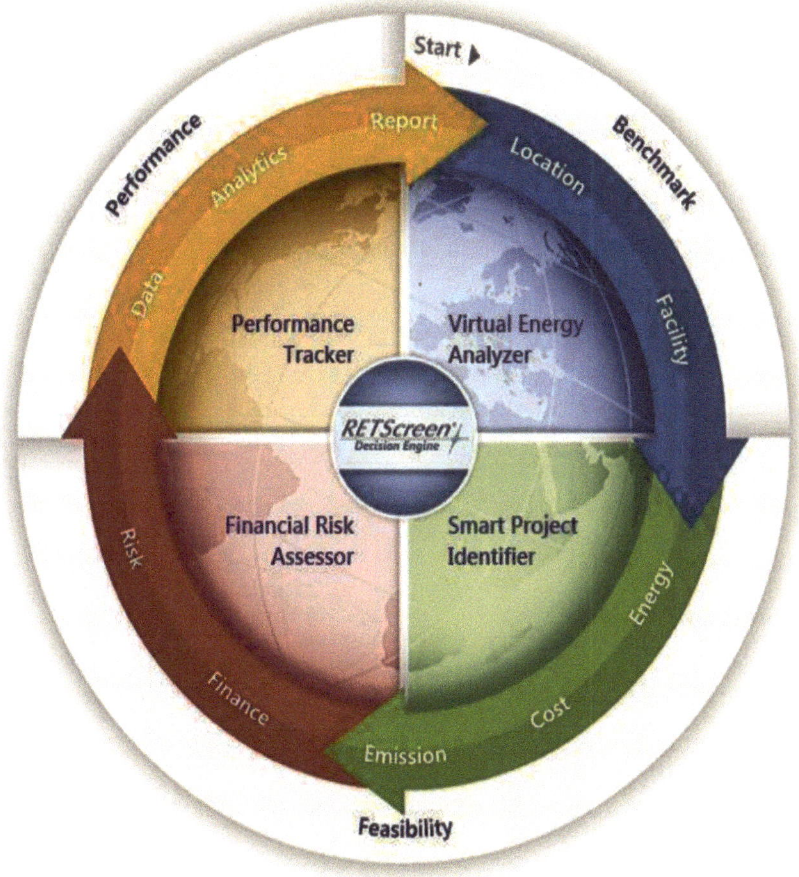

Figure 2.2 Selected research methodology to solve HRE benchmarking issues.

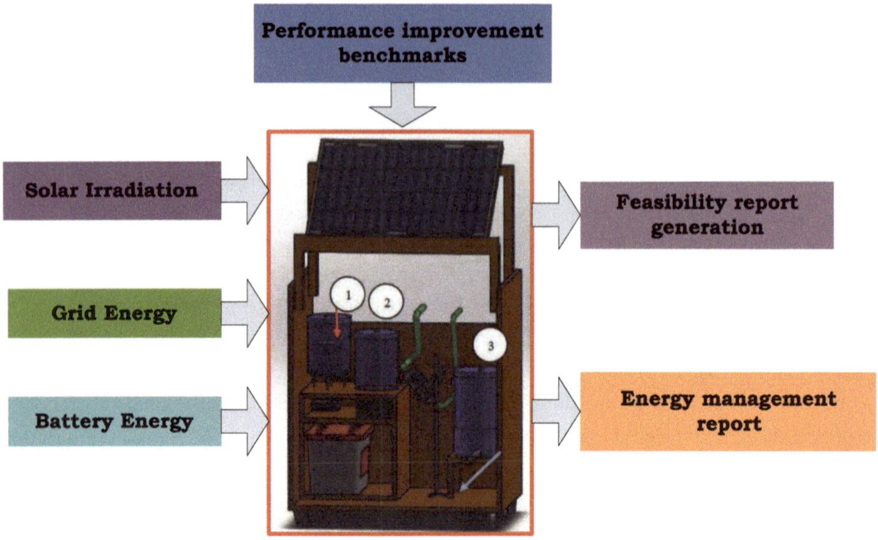

Figure 2.3 Retscreen working model for a solar water-purifier plant.

4. Results and discussion

The existing system's sustainability targets are fixed by base case performances as a first objective of a 50% utility bill reduction, second as a 70% emission mitigation, and finally a 50% increased energy use intensity process.

4.1 Fuel consumption

The existing system fuel consumption is planned to be reduced to 47%–50% to attain a significant utility bill reduction. As a primary benchmark, it aims to reduce the intensity of energy use from conventional energy sources. Therefore, the power for the water purification plant has to be fulfilled mostly by renewable energy resources. In the base case, a total energy of 1635 kWh is used for 1 year, and in the proposed case we only use 860 kWh for 1 year. Hence, the annual fuel saving is up to 776 kWh.

4.2 Emissions benchmark

This is the main target, which can have great impacts on sustainability. It is targeted at 70% within 2 years of the process beginning.

4.3 Energy cost benchmark

The energy cost is also set as a benchmark target. It is comparatively planned to reduce the utility bill to 50% of the existing system. Figs. 2.4, 2.5 and 2.6 illustrate the benchmark targets for a sustainability developed energy system. Table 2.2 illustrates all

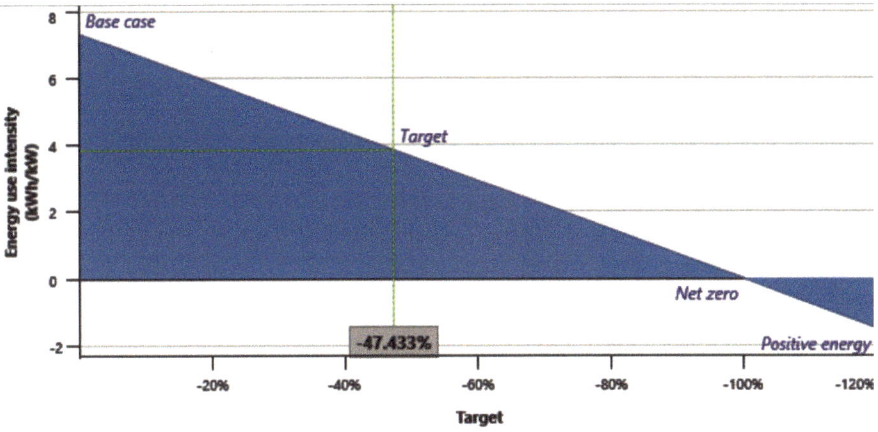

Figure 2.4 Graph for energy use intensity.

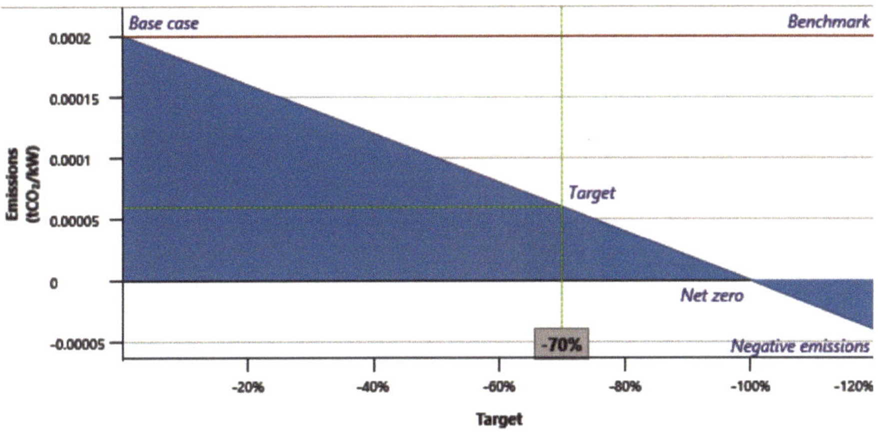

Figure 2.5 Graph for emissions.

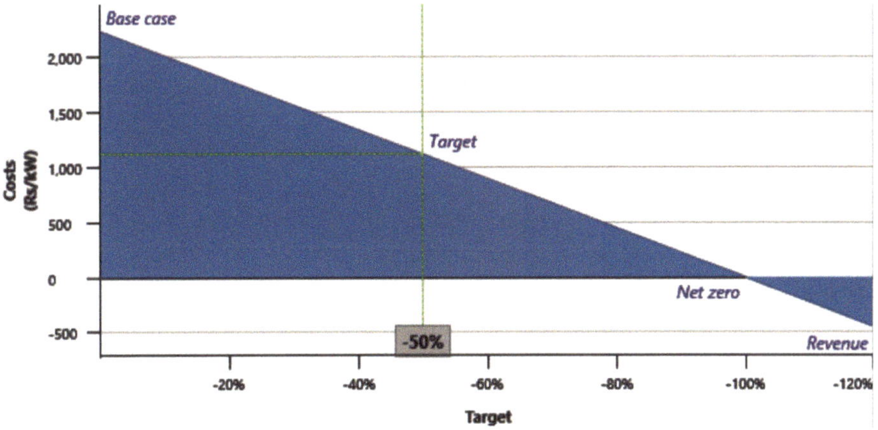

Figure 2.6 Graph for energy costs.

Table 2.2 Proposed benchmark outcomes.

	Benchmark-1 (energy use intensity)		Benchmark-2 (emissions)		Benchmark-3 (energy costs)	
Minimum−average	134	kWh/kW	0.0001	tCO_2/kW	1340	Rs/kW
Maximum−average	223	kWh/kW	0.0002	tCO_2/kW	2230	Rs/kW
Base case	7.3	kWh/kW	0.0002	tCO_2/kW	2230	Rs/kW
Reference year	2021		2021		2021	
Set target						
Year	2023		2023		2023	
Target	−47.40%		−70%		−50%	
Proposed case	3.8	kWh/kW	0.0001	tCO_2/kW	1115	Rs/kW
Facility−annual plan						
	Fuel consumption	**Units**	**Emissions**	**Units**	**Energy cost**	**Units**
Base case	1635	kWh	0.04	tCO_2	499,520	Rs
Proposed case	860	kWh	0.01	tCO_2	249,760	Rs
Fuel saved	776	kWh	0.03	tCO_2	249,760	Rs

benchmark targets and predicted outcomes in terms of annual energy use intensity, emissions, and energy cost savings after 2 years of operating time. Nearly 776 kWh of energy is obtained from solar renewable energy as a planned benchmark attainment. A reduction of 0.03% in the pollution level was obtained due to the emission mitigation. Energy costs were reduced by Rs. 249,760 by the proposed level of benchmark fixations.

4.4 Sustainability development indices

The proposed system design and improved sustainability index with the proposed model results in reduced fuel consumption which leads to a fuel saving of 1214 L compared with the existing operation. The equivalent incurred fuel cost saving is Rs. 24,273 and its corresponding GMG emission mitigation is 1.1 tCO_2 (Fig. 2.7).

4.5 Financial viability

The financial viability of the above system has been examined with necessary key measurements considering the inflation rate as 2%, assumed project life span as 20 years, debit ratio as 70%, debit interest rate assumed as 7%, and the total debit term as 15 years. The corresponding outcomes related to the proposed system's financial viability as pretax IRR equity was 323% and the pretax IRR assets were 98.4%. The investment payback as a simple payback method is 0.98 years and the equity payback method is 0.31 years, with its distribution pretax INR equity as illustrated in Fig. 2.8.

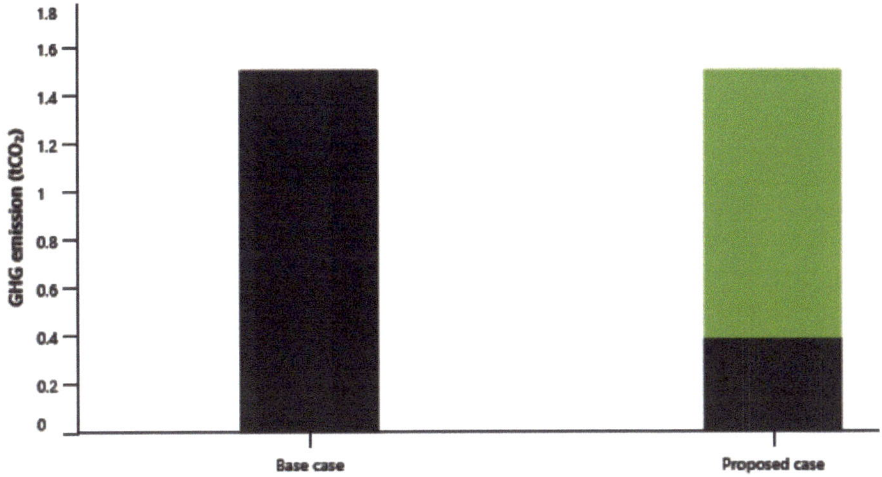

Figure 2.7 Gross annual GHG emission reduction (1.1 tCO_2 is equivalent to 0.2 cars and light trucks not used).

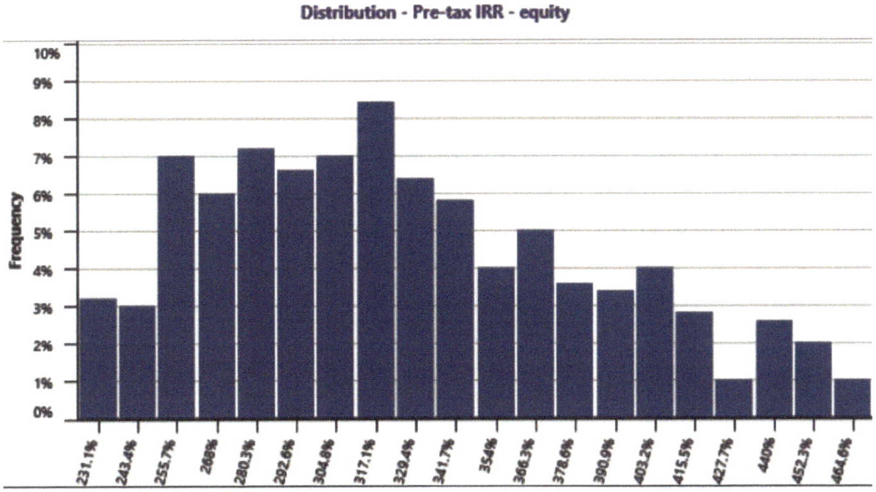

Figure 2.8 The distribution pretax IRR equity.

4.6 *Cash flow*

The proposed system's cash flows over a life span have been expressed in Fig. 2.9. This parameter gives a guideline to investors about the annual cash flow. The cumulative cash flows are expressed in Fig. 2.10.

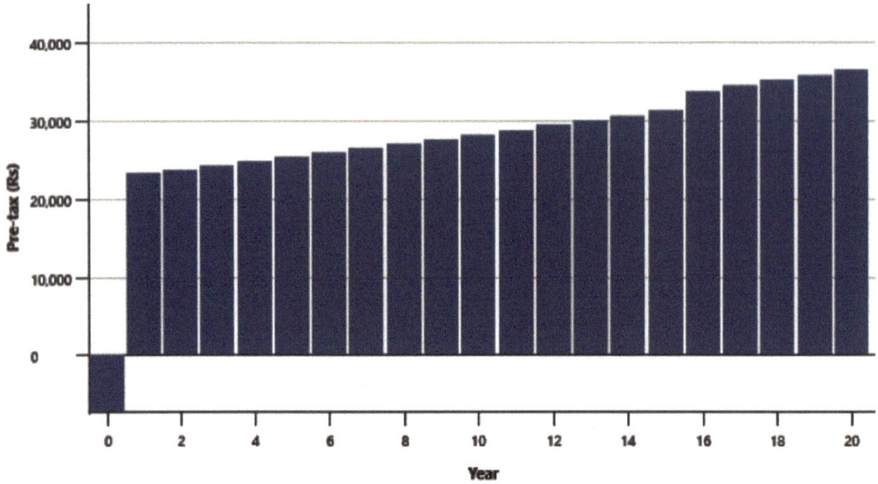

Figure 2.9 Annual cash flow of the proposal.

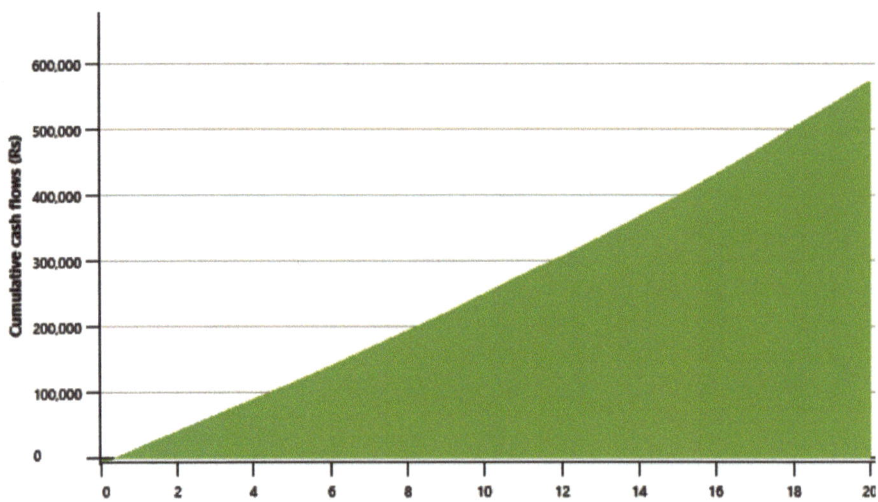

Figure 2.10 Cumulative cash flow of the proposal.

4.7 Financial risk impacts

The solar-based electrified water purification plant's financial risk and impacts are illustrated in Fig. 2.11. The debit ratio of the system will have a high positive impact. Next, the order of impacts occurs at the fuel cost. The debt interest rate impacted in the least order of the system is accepted as the level of risk.

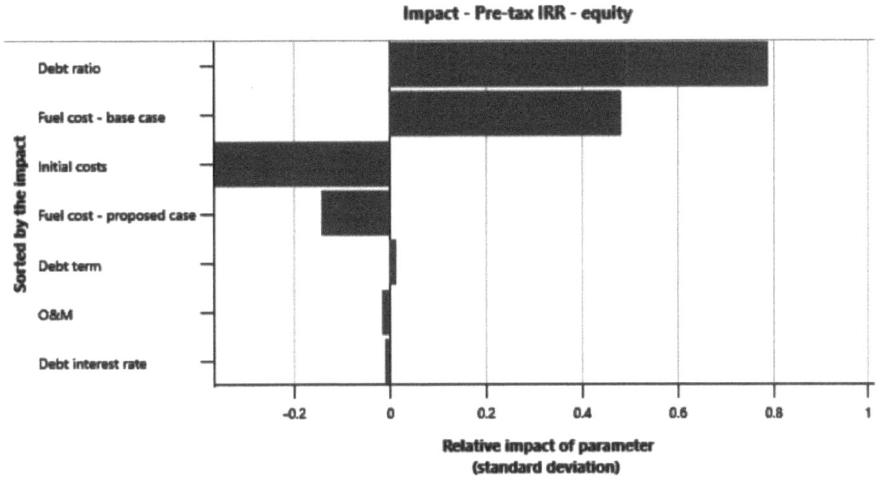

Figure 2.11 Financial risk impacts of the system.

5. Conclusion

The proposed hybrid renewable-based electrification system for water purification application described in this chapter has high technical performance with viable economic benefits as observed over the existing base case of conventional-based electrification. This case study clearly describes the techno-economic analysis of a renewable energy-based water purification system at the Hyderabad-based educational institute campus. The outcomes of this study attained sustainable energy that was able to reduce the utility energy bill and environmental pollutants. The system performance was also verified with various user constraints and accepted benchmarking values as guided by global sustainability agencies.

References

[1] W. Yamaka, R. Phadkantha, P. Rakpho, Economic and energy impacts on greenhouse gas emissions: a case study of China and the USA, Energy Reports 7 (2021) 240–247.

[2] N. Das, A. Mandal, S. Mandal, Polluted waters of the reclaimed islands of Indian Sundarban promote more greenhouse gas emissions from mangrove ecosystem, Stochastic Environmental Research and Risk Assessment 36 (2021) 1277–1288, https://doi.org/10.1007/s00477-021-02135-5.

[3] M. Krishnamoorthy, P.A.D. Vimal Raj, S. Suresh, K. Natarajan, Techno economic analysis of hybrid renewable electrification system in different climatic zones, in: T. Hitendra Sarma, et al. (Eds.), Emerging Trends in Electrical, Communications, and Information Technologies, Lecture Notes in Electrical Engineering, vol. 569, 2020.

[4] F. Alcon, J.A. Zabala, J.M. Martínez-Paz, Assessment of social demand heterogeneity to inform agricultural diffuse pollution mitigation policies, Ecological Economics 191 (2022) 107216.
[5] K. Murugaperumal, S. Srinivasn, G.R.K.D. Satya Prasad, Optimum design of hybrid renewable energy system through load forecasting and different operating strategies for rural electrification, Sustainable Energy Technologies and Assessments 37 (2020) 100613.
[6] N. Karuppiah, B. Praveen Kumar, V. Sampath, Fault detection of solar PV system using SVM and thermal image processing, International Journal of Renewable Energy Resources 10 (2) (2020) 967–977.
[7] A.B. Owolabi, K.N. Benyoh Emmanuel, W.R. Jong, S. Dongjun, H. Jeung-Soo, Measurement and verification analysis on the energy performance of a retrofit residential building after energy efficiency measures using RETScreen Expert, Alexandria Engineering Journal 59 (6) (2020) 4643–4657.
[8] A.B. Owolabi, K.N. Benyoh Emmanuel, W.R. Jong, S. Dongjun, H. Jeung-Soo, Validating the techno-economic and environmental sustainability of solar PV technology in Nigeria using RETScreen Experts to assess its viability, Sustainable Energy Technologies and Assessments 36 (2019) 100542.
[9] D. Moya, P. Juan, K. Prasad, Technical, financial, economic and environmental pre-feasibility study of geothermal power plants by RETScreen–Ecuador's case study, Renewable and Sustainable Energy Reviews 92 (2018) 628–637.
[10] K.-H. Lee, D.-W. Lee, N.-C. Baek, H.-M. Kwon, C.-J. Lee, Preliminary determination of optimal size for renewable energy resources in buildings using RETScreen, Energy 47 (1) (2012) 83–96.
[11] S. Weida, S. Kumar, R. Madlener, Financial viability of grid-connected solar PV and wind power systems in Germany, Energy Procedia 106 (2016) 35–45.
[12] M.S. Ramli, S.S.A. Wahid, K.K. Hassan, A comparison of renewable energy technologies using two simulation softwares: HOMER and RETScreen, in: AIP Conference Proceedings, vol. 1875, no. 1, AIP Publishing LLC, 2017, p. 030013.
[13] A.J. Aristizábal, C. Arturo Páez, Power quality analysis of 6 kW building-integrated photovoltaic system and economic evaluation using RETScreen, International Journal of Ambient Energy 39 (8) (2018) 813–822.
[14] K. Dikgale, D.F. Ntobela, B.G.V. Mendes, L.K. Tartibu, T.J. Kunene, E. Bakaya-Kyahurwa, Development of solar-powered water purification systems, in: 9th International Conference on Appropriate Technology. November 2020, Tshwane University of Technology, Pretoria, 2020, pp. 900–919.
[15] G.N. Nur, M. Ahnaf Sadat, Design and construction of solar water purifier, in: International Conference on Mechanical, Industrial and Materials Engineering, RUET, Rajshahi, Bangladesh, 2017.
[16] A.I. Wibowo, K.-C. Chang, Solar energy-based water treatment system applicable to the remote areas: case of Indonesia, Journal of Water, Sanitation and Hygiene for Development 10 (2) (2020) 347–356.
[17] S. Prakash, D. Toppo, Solar energy based water purification system, International Journal of Pure and Applied Mathematics 119 (12) (2018) 7863–7873.

Biomass energy conversion through pyrolysis: A ray of hope for the current energy crisis

3

R. Sakthivel[1], *G.V. Harshini*[1], *Musunuri Shanmukha Vardhan*[1], *Abhiram Vinod*[1] *and K. Gomathi*[2]
[1]Department of Mechanical Engineering, Amrita School of Engineering, Amrita Vishwa Vidyapeetham, Coimbatore, India; [2]Department of Mechatronics Engineering, Kongu Engineering College, Perundurai, India

1. Introduction

The demand for energy is constantly increasing, but the natural resources which are used to fulfill this demand are decreasing at a very high rate. Even though natural resources can be replenished it would take about hundreds of thousands of years to achieve the restoration of natural resources. Therefore, in this current scenario, we need an efficient and eco-friendly solution to meet this demand. The solution to the upcoming energy crisis is to adapt to renewable energy. Renewable energy refers to solar energy, wind power, hydroelectric power, geothermal power, tidal power, and biomass (biofuels). Therefore, biomass has a very promising future in the field of energy production. Biomass includes animal or plant extracts which can be further processed to obtain biofuel and used to generate heat and electricity. Biomass was the primary method used by early humans to extract energy, i.e.,the use of wood logs. Biomass can be classified in many forms based on the selection criteria, such as physical state, type of production used for extraction, and purpose of the final product [1]. Energy crops, agriculture crop residues, forest residue, processed wood residue, algae, municipal waste, etc., are some examples of biomasses. Some of the benefits of biomass usage are a reduction the release of GHGs (greenhouse gases), a decrease in the dependence on fossil fuels, profitable waste management, and new opportunities in agriculture and agriculture-based sectors.

Biofuels which are extracted from biomass are classified into three types, which are first-generation, second-generation, and third-generation biofuels. Sugar, starch, and vegetable oils produce first-generation biofuels. First-generation biofuels are also known as conventional biofuels. When wood, food waste, and specified biomass crops are used as biomass for extracting biofuels, they are described as second-generation biofuels. The biomass of second-generation biofuels undergoes pretreatment in which lignin is broken down. Algal feedstock is used for obtaining third-generation biofuels. The production of biofuels for first-generation and second-generation biomass after pretreatment includes the processes of transesterification, distillation, and

Green Energy Systems. https://doi.org/10.1016/B978-0-323-95108-1.00006-9

fermentation. Fermentation is used for biomasses such as sugar and starch. The primary products of fermentation are ethanol, propanol, and butanol. Transesterification is a method in which biodiesel is obtained from plant and animal fats, with distillation used to separate the byproducts. The production of second-generation biofuels can be further classified into three types on the basis of the type of biomass being used, such as homogeneous, quasihomogeneous, and nonhomogeneous. Homogeneous refers to biomass such as wood cuttings and wood chips. Quasihomogeneous includes the entire forest and agriculture residue. Nonhomogeneous includes municipality solid wastes which are low-value feedstock. In comparison to the first-generation biofuels feedstock, second-generation biofuels feedstock is cheaper. Second-generation biofuels feedstock must go through pretreatment to release the trapped sugars for the production of biofuels [2].

The third-generation biofuels feedstock consists of microalgae, animal fat, and fish oil. Third-generation biofuels have an advantage compared to the other two biofuels in availability, adaptability to climatic conditions, effects on the food chain, and economic feasibility. This can be mainly attributed to the fact that algae can be cultivated in harsh conditions and the cultivation does not occupy as much space compared to traditional farmland used for first-generation biofuels feedstock. It also has a lipid content of 90% when the conditions are favorable. Algae are used to extract biodiesels, which are generally methyl or ethyl esters. This is done by a method called transesterification in which the triglycerides from the feedstock and alcohols such as ethanol or methanol are reacted in the presence of a catalyst to give the required fatty acid esters, and glycerol is a byproduct. Another method for biofuel extraction from algae is fermentation, however in this process the biofuel obtained is bioethanol.

There are many advantages to biofuels. For instance, if we consider ethanol and compare them with conventional gasoline, ethanol burns cleaner than gasoline and produces lower GHGs. When blended with fuels, it increases the octane number (a measure of the ability of a fuel to resist knocking when ignited in an IC engine). Along the same lines, biodiesels are also less polluting than their counterpart, i.e., fossil fuels diesel, and have good blending properties. Due to the lubricating properties of biodiesels, they also tend to increase the life of diesel engines. As for the applications of biofuels in the automobile sector, the usage of biofuels has been already initiated. Many developed countries have already looked into their policies with a wide range of uses of biofuels either as a blend or directly to reduce their GHG emissions. Biofuel use in the automation sector is not a recent breakthrough. The Ford Model T (1908) was designed to operate on gasoline and a low level of ethanol. E85 is a blend of 85% ethanol and 15% gasoline; this blend can provide more torque, greater horse power, and even has a cooling effect on the engine [3]. It is safe to say that there will be an increase in demand for biofuels in the future.

Biomass can be converted into different forms of energies using conversion techniques. There are three main technologies which the conversion biomass goes through: these are thermochemical, biochemical, or biological and mechanical extraction. Techniques like combustion, pyrolysis, gasification, and liquification come under thermochemical processes, whereas digestion and fermentation come under biological processes [4]. Fermentation is a process in which biomass is converted into ethanol.

In fermentation, microorganisms such as yeast break down the sugars in the organic materials to form alcohols or acids [5]. Therefore, any material with a high sugar content can be used as biofuel feedstock. Anaerobic digestion (absence of oxygen) converts the biomass into renewable natural gas.

Gasification is a thermochemical process in which all carbonaceous matter, solid or liquid, is converted into syngas. Syngas consists of CO and H_2. There are huge amounts of heat released during this process [6]. The temperature range of this process is between 600 and 1500°C. Plasma gasification, melting gasification, and microwave gasification are some of the most commonly used gasification technologies in industries. Liquification is a process in which the solid biopolymeric structure of the biomass is broken down into liquid state using hot and pressurized water for a certain amount of time to form biofuels as the end-product. The typical conditions of the liquification process are 533 and 647 K temperature and a pressure of 4–22 MPa. Depending on the severity of the process the hydrothermal process is divided into three types, namely hydrothermal carbonization, hydrothermal liquification, and hydrothermal gasification [7].

In general, pyrolysis is a chemical process which takes place in elevated temperatures in an inert atmosphere. Pyrolysis is a thermochemical process in which organic matter is decomposed in the absence of oxygen at elevated temperature. The pyrolysis process is a complex process in which successive reactions take place where organic material is heated in a nonreactive atmosphere. In this conversion process, thermal disintegration of molecules takes place yielding biochar, bio-oil, and syngas. The advantages of pyrolysis techniques include that they can be performed at smaller scales and even in remote locations, enhance the energy density of the biomass, and reduce transport and handling costs. The pyrolysis process is an endothermic process and a sufficient amount of heat needs to be provided to complete the process.

There can be wide range of feedstock used for pyrolysis. Feedstock of pyrolysis depends upon the moisture content and particle size mostly. The recommended moisture content for a pyrolysis feedstock is around 10%. At higher moisture contents, there is a risk of producing dust instead of oil. High moisture content wastes, like sludge and meat waste, are dried before pyrolysis processing. Pyrolysis techniques can process particles only at a maximum of 2 mm for rapid heat transfer through particles. Therefore, the feedstock particles are reduced in size for processing via pyrolysis. Pyrolysis techniques are further classified on the basis of operating temperatures, namely slow pyrolysis (300–550°C), fast pyrolysis (500–600°C), and flash pyrolysis (900–1300°C). The residence time is the length of time during which a substance is in a given condition. It is different for each type of pyrolysis, the highest being for slow pyrolysis, with the residence time being 2–30 min, the fast pyrolysis residence time is about 0.5–2 s, and it is less than 0.5 s for flash pyrolysis. Typical operating parameters of pyrolysis are given in Table 3.1.

The product yield of the slow pyrolysis process is 30% oil, 35% char, and 35% gas. Due to the low production of bio-oil, slow pyrolysis is unlikely to be used as this can be attributed to the high residence time which affects the yield. The product yield in fast pyrolysis is much more appreciable than that of slow pyrolysis, and is about 60% bio-oil, 20% biochar, and 10%–20% gas. High heat transfer and high heat rates,

Table 3.1 Typical operating parameters for pyrolysis.

Pyrolysis process	Residence time (s)	Heating rate (K/s)	Particle size (mm)	Temperature (K)
Slow	450–550	0.1–1	5–50	550–950
Fast	0.5–10	10–200	<1	850–1250
Flash	<0.5	>1000	<0.2	1050–1300

very short vapor residence time, rapid cooling vapors, and precision control of reaction temperature are some of the characteristics of fast pyrolysis. Fast pyrolysis is incredibly popular because of its range of specialty and commodity chemicals. The fast pyrolysis technologies have a relatively low investment, and high-energy efficiencies especially for small-scale operations. Flash pyrolysis has the highest bio-oil yield production of 75%. Flash pyrolysis can be characterized as quick devolatilization (emission of volatile gases) in an inert atmosphere [8]. Reactors are the most important part in a pyrolysis reaction. Particle size and vapor residence time have the least effect on the bio-oil yield, but these factors affect the composition of bio-oil. There are many types of reactors available at disposal, but each has its own advantages and disadvantages. A fixed-bed reactor is an example of a reactor which has a gas cooling and a cleaning system. They are considered for small-scale heat and power applications. Some other examples of pyrolysis reactors are fluidized bed reactors, bubbling fluidized bed reactors, vortex reactors, ablative reactors, vacuum pyrolysis reactors, rotating cone reactors, etc.

This chapter initially discusses the product mechanism in biomass pyrolysis and different types of reactors which are used in pyrolysis. As inferred initially, pretreatment is also required for biomass pyrolysis. Physical as well as chemical pretreatment techniques are discussed after different types of reactors. The influence of temperature, heating rate, and residence time on the product reaction is covered following the pretreatment techniques. Modeling and advancements of biomass pyrolysis are discussed before the conclusion.

2. Biomass pyrolysis—mechanism of product conversion

Pyrolysis is a complex thermochemical conversion process that involves several chemical reactions due to which the course of overall product formation is still unpredictable. Some of the most common chemical reactions observed in typical biomass pyrolysis are charring, isomerization, dehydration, condensation, depolymerization, aromatization, and decarboxylation. These groups of reactions can be broadly categorized into two types, i.e., primary and secondary reactions. The primary reaction zone includes fragmentation where the covalent bonds of monomers and polymers are fragmented. During fragmentation, the short-chain volatile gases and traces of noncondensable gases are produced. Next, the major domination was done by a depolymerization reaction which produced volatiles and gases. Along with this, the combination or condensation of aromatic benzene rings results in the formation of polycyclic carbon through a charring reaction. On the other hand, secondary reactions occur from the products of a primary reaction due to its instability. The major secondary reactions include cracking and recombination, where the former is responsible for the formation of lighter compounds and the latter results in heavier compounds. Biomass constitutes three major building blocks, namely cellulose, hemicellulose, and lignin. The products formed during pyrolysis stem from any of these components, due to which study of the reaction mechanism becomes inevitable.

2.1 Conversion of cellulose

Cellulose is the primary cell wall component, containing a linear polysaccharide of monomers called cellobiase, which also contains two glucopyranese bridged with glycosidic bonds. The interaction of three hydroxyl groups in the cellulose gives it a crystalline structure, stability, and strength. The temperature of cellulose degradation ranges between 315 and 400°C. With the higher degree of the polymerization reaction with suitable conditions, cellulose can reach up to 5000 units. The cellulose pyrolysis yields major products like char, water, acids, gases, anhydrosugars, and alcohols. The preliminary kinetic model for cellulose degradation through pyrolysis has been established by Bradbury et al. and is as follows [9]:

Cellulose —→Active cellulose

Activated cellulose—→Char + Gas

Activated cellulose—→Tar—→Gas

Several literature works have been reported to detail the mechanism of cellulose conversion [10–13]. Most research work has described the Diebold model as the best for a generalized pyrolysis process of biomass, whereas it does not fit well for secondary product evaluation and high heating rate studies. Meanwhile, the Broido–Shafizadeh model, which is obtained from low heating rate studies, predicts that biomass pyrolysis will occur at a more rapid rate than the observed limit. Lin et al. [14] utilized TGA and reactor setup to analyze the mechanism of cellulose pyrolysis. These authors observed the formation of levoglucosan from cellulose depolymerization as the first step, followed by the dehydration and isomerization reactions to form anhydrosugars. These sugar molecules formed furan in subsequent reactions. In addition, carbon dioxide is formed due to the decarbonylation and decarboxylation reactions. Finally, char is also produced as the byproduct of cellulose degradation. Wooten et al. [15] offered a novel model for cellulose pyrolysis at low temperature around 300–350°C using NMR studies of products. Patwardhan et al. [16] investigated the secondary reaction mechanism of cellulose pyrolysis using two different reactor setups. These authors found that oligomerization of levoglucosan and decomposition of primary products constitutes the secondary reaction in cellulose pyrolysis.

From the above discussions, it is clear that cellulose pyrolysis at lower temperature (<350°C) favors the formation of water, char, and gases owing to the dehydration reaction. At intermediate temperatures (up to 450°C), the formation of levoglucosan predominates. It is worth noting that major conversion of cellulose takes place between 300 and 390°C, whereas above 400°C the aromatic residue is formed.

2.2 Conversion of hemicellulose

Hemicellulose is the heterogeneous cluster of branched polysaccharides which acts as a connecting link between cellulose and lignin. It comprises structural elements like mannose, arabinose, glucose, glucuronic acid, xylose, and galactose. The hemicellulose layer is bridged with cellulose through a hydrogen bond, whereas the lignin linkage is made by an ester bond. The decomposition temperature of hemicellulose ranges from 220 to 350°C. Around 25–35 wt. % of hemicellulose can be observed in a typical dried woody biomass sample. During a normal pyrolysis procedure, hemicellulose is degraded similar It to cellulose. The reactions were observed to be similar at a given temperature range, whereas at 200–260°C more volatiles were traced in hemicellulose decomposition. Meanwhile, a higher concentration of acetic acid along with low levoglucosan was generated during cracking of the hemicellulose layer in the biomass sample. The staged kinetic model is one of the first proposed models for hemicellulose pyrolysis followed by a multistage model [17,18]. The multistage model enunciates that formation of an active hemicellulose component as the initial step followed by its decomposition to form char and gases by parallel reaction pathways. After several research outcomes, xylan also has been selected as the chemical component to predict the reaction pathways in pyrolysis of hemicellulose. It was observed that weight loss of xylan occurred at a temperature range of 220–320°C through an exothermic reaction. The products generated include a range of ketones, acids, aldehydes along with gases like methane, hydrogen, and carbon dioxide. It can be seen that at the above decomposition range (>350°C), the char production dominates the volatile generation and constitutes the main reaction. Dehydration and bond breaking take place at 150–270°C, whereas polymerization takes place above 250°C until the maximum decomposition range [11,19].

2.3 Conversion of lignin

Lignin is the aromatic, cross-linked phenolic polymer responsible for the structural rigidity of the cell wall through by the polysaccharide fibers together. It is made up of phenyl propane monomers such as guaiacyl, syringyl, and p-hydroxyphenyl units. The decomposition temperature is greater than that of cellulose and hemicellulose, which ranges between 160 and 900°C. The composition of lignin is varied based on the type of biomass such as softwood (23–33 wt.%) and hardwood (16–25 wt.%) [20]. The heterogeneous chemical structure of lignin is comprised of various bonds like ether (C–O), C–C bonds, –OH bonds, etc. TGA analysis revealed that primary pyrolysis reactions occur at 350°C, while the methoxy group reactions can be observed at a range of 400–450°C. The secondary reactions related to catechol gasification can be seen in the 550–600°C range [21]. Lignin pyrolysis yields a variety of solid, liquid, and gaseous products. The liquid products mainly constitute phenolic compounds, whereas some gases like hydrogen, oxides of carbon, hydrocarbons, and aldehydes are also produced. On the solid side, char is the major aromatic structure with half of the energy content of native biomass. The char is generally formed by polymerization of lower radical species along with removal of side function group compounds. Lignin pyrolysis starts with water evaporation followed by the evolving of major volatiles. Finally, the reaction ceases with the formation of lower molecular weight gaseous compounds. The increase in pyrolysis temperature leads to effective breaking of aliphatic −OH, −C, and −CH bonds.

Char formation during the pyrolysis of lignin in biomass is greatly influenced by the amount of methoxyl functional groups. Some of the major compounds detected during pyrolysis of different types of wood are derivatives of syringyl and guaiacyl, along with their respective aldehydes and alcohols. The mechanism of fast pyrolysis of lignin has been studied by Custodis et al. using a microreactor setup. These authors observed that radical fragment decomposition was in accordance with Franck–Condon simulation [22]. It is also to be noted that the product quality and distribution of biomass pyrolysis strongly depends on the lignin origin and its orientation. Phenols are the major product of lignin pyrolysis irrespective of the feedstock used, due to breaking of ether and C–C bonds. As compared to cellulose, lignin pyrolysis yields greater amounts of phenols and char. Finally, as far as the reactions are concerned, alkyl chain breaking occurs at 150–420°C, whereas the conversion of an aromatic ring and production of char occurs between the temperature range of 380 and 900°C.

3. Biomass pyrolysis reactors

The reactor of a biomass plant plays a very important role in biomass production. It functions as the powerhouse of the plant. The major part of the reactor is the heating system. Various types of reactors have been found and tested over the past decade. The efficiency of the reactor is determined based on the quality of the product and the production time. Research in these fields has resulted in the design of several biomass reactors. These reactors can be classified based on the type of reaction that takes place in the reactor. Based on the type of reaction, the reactors can be classified into two types: slow pyrolysis reactors and fast pyrolysis reactors. The subdivisions of the reactors based on the types of reactors are represented graphically in Fig. 3.1.

3.1 Fast pyrolysis reactors

To maximize the production of bio-oil in a short time, various methods have been designed with different advantages and disadvantages. Some of these types of reactors are discussed below.

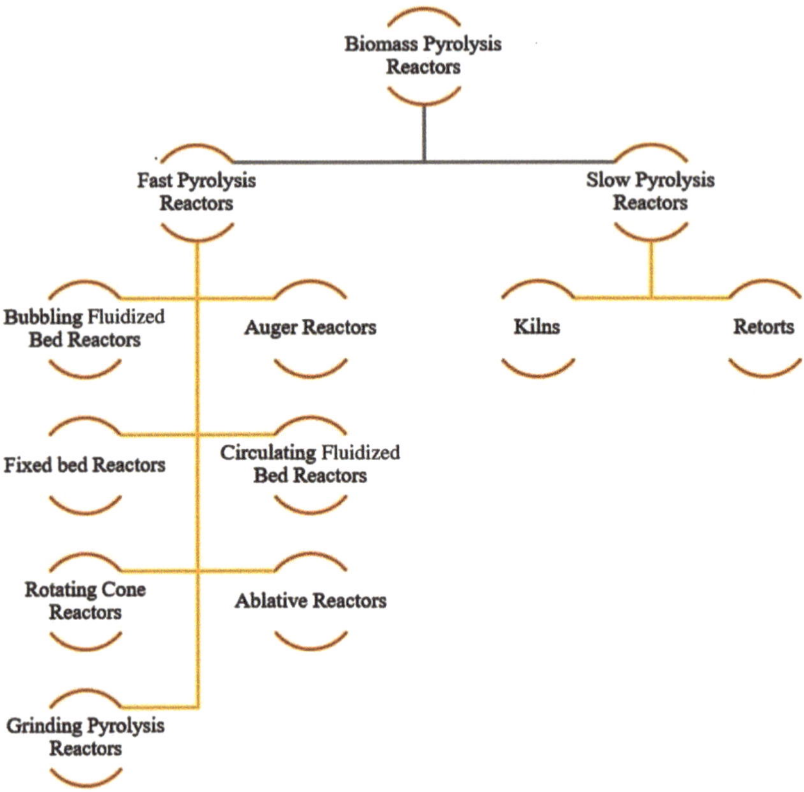

Figure 3.1 Classification of biomass pyrolysis reactors.

3.1.1 Bubbling fluidized bed reactors

Bubbling fluidized bed reactors have a yield of about 70%–75% on a dry feed basis of wood. The major requirement for a high yield from the reactor is the size of the feed of the reactor. For a high yield, the size of the feed should be close to 2–3 mm [23]. The size determines the heating rate of the reactor. The uniqueness of this reactor is the fluidized flow of gases, which helps in controlling the residence rate of the solids and the vapors. After the residence of the char takes place, it is separated from the bio-oil using an ejection and entrainment system followed by complete separation in a cyclone. The number of cyclones can vary based on the product requirements. The structure of a bubbling fluidized bed is represented in Fig. 3.2.

The research in bubbling fluidized bed reactors was initiated at the University of Waterloo in Canada [23]. Later research studies in the same field were conducted in the United Kingdom, where a 250 kg/h pilot plant was constructed. To improve the yield of the plant, various research studies have been conducted. The yield of the reactor depends on various factors such as the type of feed and the size of the feed. In research conducted by Ali et al., the operational parameters were studied for the pyrolysis product using maize stock as feed at an operating temperature of 360–540°C,

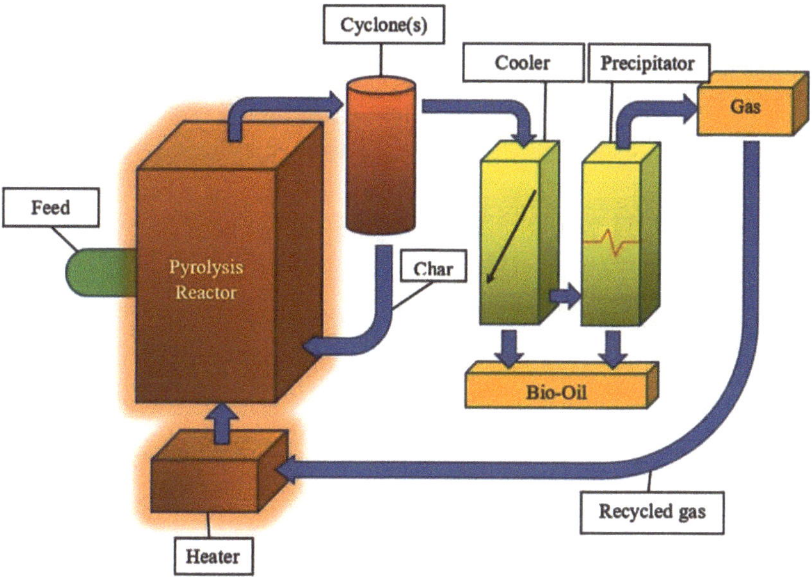

Figure 3.2 Structure of a bubbling fluidized bed reactor.

feed size of 1–2 mm, and a gas flow rate of 7–13 m^3/h. The maximum yield achieved in the experiment was 42 wt.% at particular operational parameters [24]. Seok et al. concluded that the heat transfer that takes place during the reaction is governed by the bubbles formed during the reaction. This was concluded by using computational fluid dynamic simulations of a reactor that uses wood as a feed.

3.1.2 Auger reactors

The design of the auger reactor consists of a long cylindrical tube with a rotating auger inside placed along the axis of the cylindrical case. The tube is oxygen-free. Hot sand and biomass are used as input to the reactor, where the reaction takes place once the optimum temperature is achieved. In this reactor, the hot sand acts as a source of heat for the reaction. The gases that are produced from the reactor are collected at the top and are supplied to a condenser where the vapor condenses. The sand and the char produced are removed from the reactor with the aid of the rotating auger. These reactors are classified into three types based on the scale of the reactor. The classifications are laboratory-scale reactors (feed rate <1 kg/h), pilot-scale reactors (feed rate 1–15 kg/h), and industrial-scale reactors (feed rate >15 kg/h) [23]. The structure of an auger reactor is presented in Fig. 3.3 [25].

Various research studies have been conducted on auger reactors by varying the type of feed, temperature, and feed rate to study their effects on the yield of char and bio-oil [23]. Yu et al. studied the effect of different temperatures on the yield of reactors. Two types of feed (rice husk and corn stock) were used for the study at temperatures of 350, 400, 450, 500, 550, and 600°C. The maximum yield was observed at 500°C for both

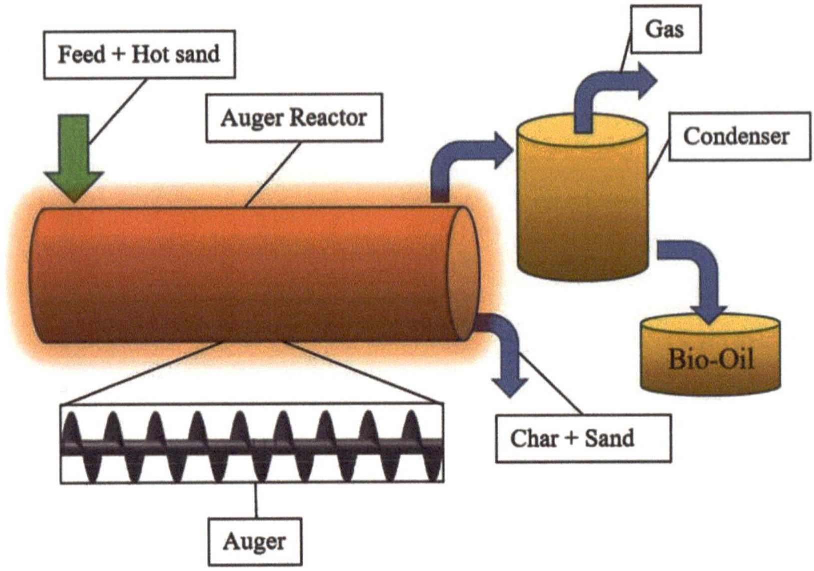

Figure 3.3 Structure of an auger reactor.

types of feed. The reactor showed a yield of 51 wt.% when rice husk was used as feed and 54 wt.% when using corn stock was used as feed at an optimum temperature of 500°C [26]. In the study conducted by Mathew et al., mesquite sawdust and rice straws were used as feed for a 500-L reactor. They studied the effects of process conditions such as the temperature and rpm of the auger in the reactor. The maximum bio-oil yield of 42.6 wt.% was achieved using mesquite sawdust at 500°C and 30 rpm, and a yield of 34.6 wt.% was obtained by using rice straws at 475°C at 50 rpm [27].

3.1.3 Fixed-bed reactors

Fixed-bed reactors are small-scale reactors that are mostly used for laboratory purposes due to their good controllability features. Large-scale applications have not been completely designed and tested. Research on large-scale applications such as homogeneous distribution of heat in the reactor is in complete. Xiong et al. used a fixed-bed reactor to study the yield of the reactor at different temperatures and heating rates while using rice husk as feed. Their study revealed that the yield of coke was high at low heating rates. They also observed that the formed coke was nonporous at low heating rates and at high heating rates the coke became porous [28]. The structure of a fixed-bed reactor is presented in Fig. 3.4 [29].

3.1.4 Circulating fluidized bed reactor

Circulating fluidized bed (CFB) reactors are very similar to fluidized bed reactors. The difference is that the residence time of the gases and the vapors is different. CFB reactors also produce bio-oil with a higher char content. Another unique feature of

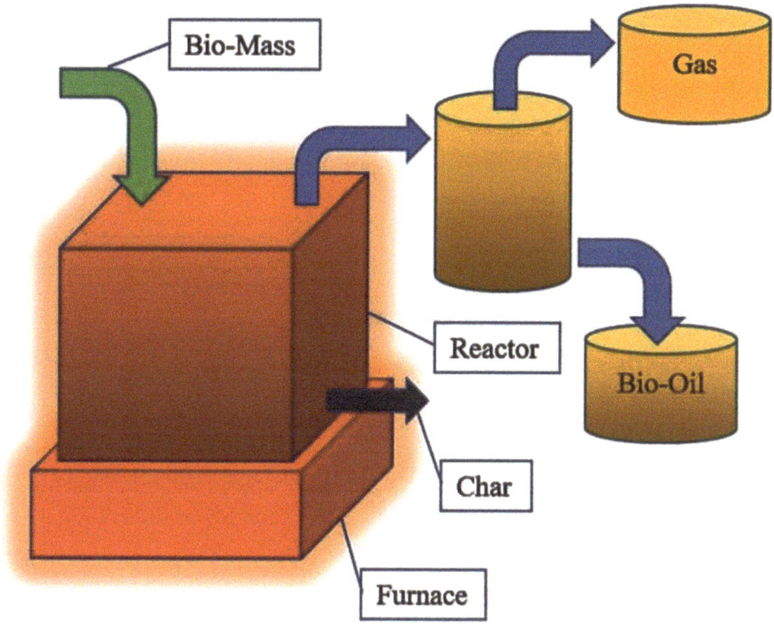

Figure 3.4 Structure of a fixed-bed reactor.

this reactor is that the heat is supplied by recirculating heated sand from a secondary char combustor. The complex hydrodynamics of the reactor does not affect the output [23]. A higher output from reactors resulted in its usage by petroleum and petrochemical industries. The structure of a CFB reactor is presented in Fig. 3.5.

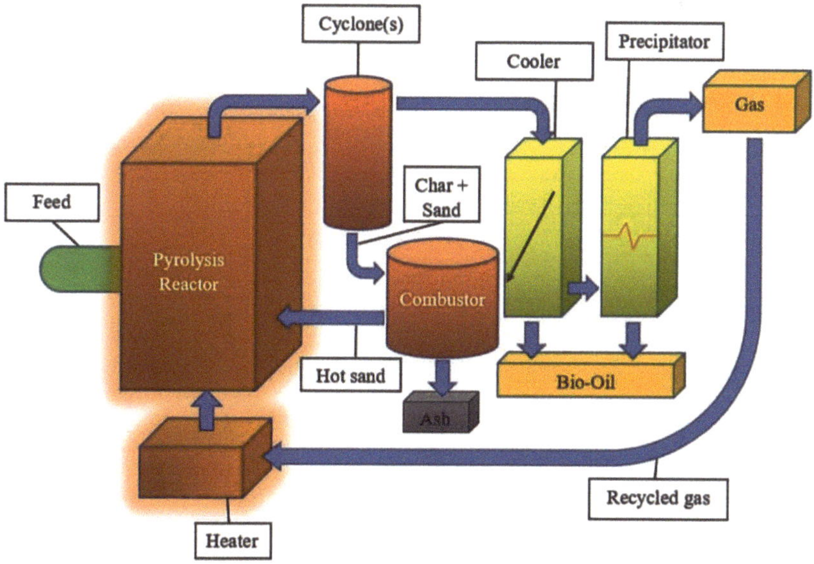

Figure 3.5 Structure of a circulating fluidized bed reactor.

Similar to bubbling bed reactors, the yield of a CFB reactor depends on the feed of the reactor. Velden et al. used biomass as a feed to study the characteristics of the reactor. The major output from the experiment was bio-oil. The maximum yield from the reactor was 60 wt.% at 500°C [23]. A similar study was conducted by Lappas et al. where the effect of ZSM-5 catalyst was also studied. The major aim of this research was to study the composition of the products from the reactor. The maximum yield of the reactor was 70 wt.% without the use of the catalyst. The study concluded that the addition of catalyst increases the emissions, such as water, coke, and gases, from the reactor compared to conventional pyrolysis in a CFB reactor without the use of a catalyst. However, the use of catalysts improved the quality of the bio-oil produced [23]. Trendewicz et al. studied the effect of different feeds for a CFB reactor. In the research conducted, the yields of different feeds such as pine, wheat straw, olive husks, and waste at a feed rate of 0.023 kg/s were analyzed. The results showed that the highest yield was obtained from pine with a yield of 65.5%, followed by waste at 61.2%, straw at 59.6%, and olive husk at 58%. Dai et al. studied the effects of temperature, feed size, and feed position in a CFB reactor using wood as the feed. From the research, they concluded that higher temperature and lower residence time produced a lower quantity of fluids [30].

3.1.5 Rotating cone reactors

To achieve fast pyrolysis of biomass, a rotating cone reactor was designed by the University of Twente. This reactor was designed to include a rotating cone which helped in mixing hot inert particles with the biomass to increase the amount of heat transferred to the biomass mixture. When compared to a fluidized bed reactor, the rotating cone reactor does not require inert gases to function. This advantage considerably reduces the number of steps required for production. The first design of the rotating cone reactor did not involve the use of inert sand. A sand transported rotating cone reactor was later developed. The structure of a rotating cone reactor is presented in Fig. 3.6 [23].

In the research conducted by Junsheng, wood shatters were used as a feed in the rotating cone reactor at 550°C and a vacuum pressure of 0.08 MPa. The yield of bio-oil from the pyrolysis was 54.83% [23]. Wagenaar et al. studied the effect of cone rotational speed, reactor volume, feed rate, and reactor temperature in a rotating cone reactor using wood dust as feed. The results from these experiments were compared with predictions from an integrated reactor model. They achieved a maximum yield of 70 wt.% [31].

3.1.6 Ablative reactors

The design of ablative reactors was aimed at reducing the cost and equipment size, and improving the controllability of the reactor. The heat is transferred to the feed by its motion over a hot surface. The effectiveness of the reactor is enabled by the amount of heat transfer that takes place before the reaction starts. The controllable factors of the reaction are the temperature, feed rate, and pressure at the hot surface. The use of inert gases becomes unnecessary due to the direct transfer of heat from the hot surface.

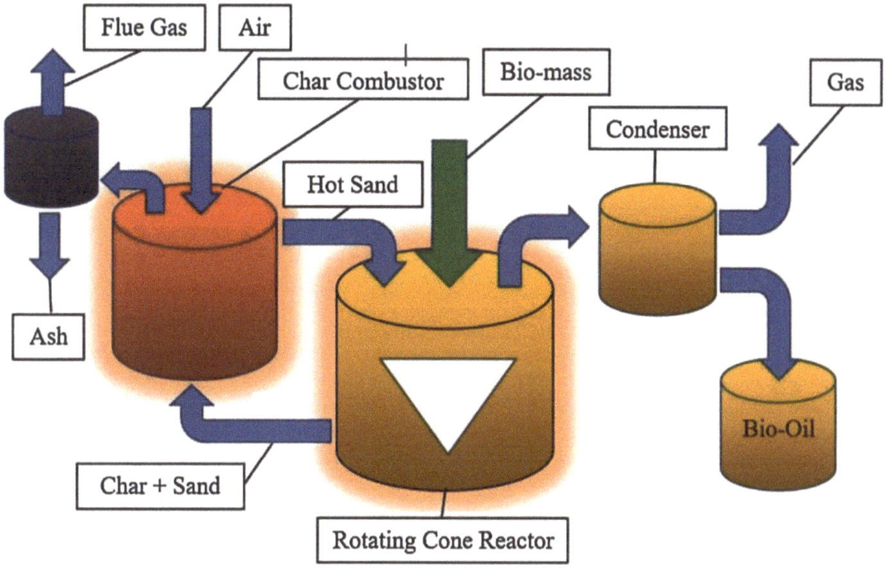

Figure 3.6 Structure of a rotating cone reactor.

Among the several reactors that have been designed, a few of them were unique due to the qualities they showcased. The National Renewable Energy Laboratory designed an ablative reactor that had a bio-oil yield of 60–65 wt.%. Aston University was able to design an ablative reactor that had a bio-oil yield of 70–75 wt.%. In Germany, ablative reactors were built to produce bio-oil which was further used to generate electricity. Various studies also have been conducted in the field. Luo et al. studied the effect of feed size by using whole wood chips and rods as feed. The results from the research showed that the yields from both feeds were very similar at around 60 wt.% [32]. Gomez-Monedero et al. studied the effect of various feed types such as *Eucalyptus* chips, camelina straw pellets, and wheat straw pellets on the yield of an ablative reactor. The results showed that eucalyptus chips produced a yield of 42.4 wt.%, camelina straw pellets a yield of 48.9 wt.%, and wheat straw pellets a yield of 41 wt.%. Another conclusion drawn from this research is that the composition of the product changed for each feed type [33].

3.1.7 Grinding pyrolysis reactors

The grinding pyrolysis reactor involves an initial step of grinding which removes the brittle outer layer of the biomass that is fed to the reactor. The structure of a grinding pyrolysis reactor is presented in Fig. 3.7 [34]. This grinding process is achieved using hot steel balls of different sizes within the reactor. The removed brittle cover forms char within the reactor, while the rest of the mass continues its reaction. Inert gases are used to push the gases out of the reactor. The advantage of the grinding pyrolysis reactor is the freedom of the input feed size. Hasan et al. studied the effects of different feed sizes on the reactor. They concluded that the reactor can use any size of feed, which could also be ground simultaneously [34].

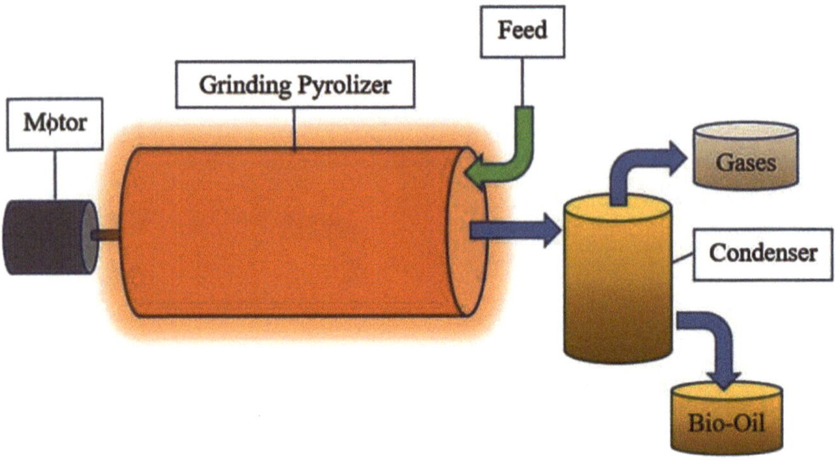

Figure 3.7 Structure of a grinding pyrolysis reactor.

3.2 Slow pyrolysis reactors

Slow pyrolysis rectors are used to produce char with reaction times of around an hour at a moderate temperature. These reactors are of multiple types. Some of these types are discussed below.

3.2.1 Kilns

Kilns are thermally insulated chambers used as heating units for various applications such as pyrolysis reactors. In a kiln, the heat required for the reaction is directly generated inside the reactor itself to ensure the direct transfer of heat to the biomass feed. There are various types of kilns based on the shape and design. The types of kilns are presented in Table 3.2 [35]. Kilns are very basic and very easy to construct. They are mostly constructed close to the biomass source to ensure a continuous supply of feed to the reactor. In a kiln, the existing biomass remains transfer the heat to the

Table 3.2 Types of kilns.

S. no.	Type of kiln	Description
1	Casamance mold kiln	- Kiln incorporated with a chimney to improve heat transfer
2	Beehive kilns	- Made from bricks in a definite structure - Improve the quality of the reaction inside
3	Missouri kilns	- Made from reinforced concrete. The concrete-fixed structures aid in reactions by insulating air gaps within the reactor walls

entering feed. This cannot be relied upon because heat will not be completely transferred and an incomplete reaction could lead to the complete burning of the feed to ashes. Another big disadvantage is that emissions from a kiln are very high.

3.2.2 Retorts

Retorts are chambers that are used to separate two fluids by distillation. Retort functions are based on heat supplied externally by different means. For biomass pyrolysis, heat is supplied to the feed by injecting hot gases into the reactor or through the walls of the reactor. Since there is no contact with the outer atmosphere, the biomass does not come into contact with oxygen, leading to higher production of char from the reactor. Various types of retorts have been designed for biomass pyrolysis to improve the production of char from the reactor. Each of these types is compared in Table 3.3 [35].

4. Pretreatment techniques

Pretreatment of feedstock is a set of processes where by the feedstock undergoes some treatment techniques before being converted into fuel (Fig. 3.8). It consists of one or more processes that produce better fuel with both good quality and quantity. General pretreatment techniques are grinding, torrefaction, and drying [36]. This pretreatment makes the path easy for the conversion of biomass into valuable products by pyrolysis.

Table 3.3 Types of retorts.

S. no.	Type of retort	Description
1	Van Marion retort	Retort with two separate reactors using a single combustion zone. Provides a charcoal yield of 30%–32%
2	Arkansas retort	Retort used steel wagons to move biomass in and out of the reactor. Heat transferred from the retort to the fixed bed in the wagons. Effective but requires a lot of maintenance
3	Reichert retort	Direct heating of biomass by passing hot vapors into the chamber. Multiple retorts are used to enable maximum efficiency. These retorts uses the steel wagons to move biomass in and out of the reactor. Heat is transferred from the retort to the fixed bed in the wagons. This type of retort is effective but requires a lot of maintenance
4	Lambiotte retort	Uses gravitational forces to transport wood to the retort. The heat from the descending charcoal is absorbed by gases which are used for heating the biomass at the top. Additional heat is provided by external means

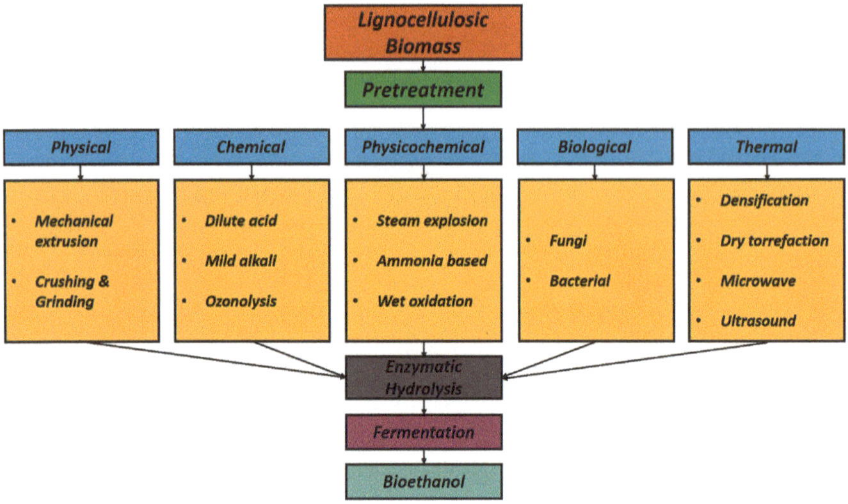

Figure 3.8 Classification of pretreatment techniques.

It enhances the process selectivity for products as it makes the biomass polymers more accessible by opening up. It opens the polymer fibers to enhance the conversion of cellulose, hemicellulose, and lignin [37]. The feedstock which is stored requires some pretreatments before its pyrolysis usually. The main aim of pretreating the biomass is to reduce and destroy the lignocellulosic structure (Fig. 3.9). This enhances the efficacy of the final pyrolyzed products. There are different techniques used for this which are discussed below.

Pretreatment can be broadly divided into five main categories:

1. Physical techniques
2. Chemical techniques
3. Physiochemical techniques
4. Biological techniques
5. Thermal techniques.

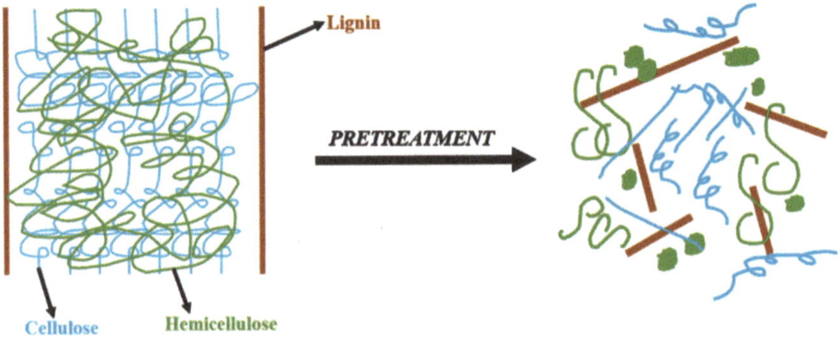

Figure 3.9 Modification of the lignocellulosic structure during pretreatment.

4.1 Physical techniques

To maintain an uninterrupted feeding system, the flowability of the feedstock into the reactor plays an important role. The flowability and material handling of the feed depend on various factors such as particle size, particle shape, density, and fluidity of the feed [37]. Feed size plays a very important role in pyrolysis reactions. A small feed size increases the flowability of the feed and thus increases the feed rate into the reactor. It also increases the reaction speed of the biomass by increasing the heat transfer rate within the reactor. Biomass is a poor conductor of heat, and thus a large feed will increase the amount of heat required to process the reaction. A smaller feed enhances the pyrolysis reaction by providing a uniform temperature distribution within the reactor. This increases the yield of bio-oil and reduces the char production from the reactor [38].

The next most important factor that affects a pyrolysis reaction is the density of the feedstock. Generally, for a smaller size of particle, the density of the feed will be high. The high density of the biomass pellets will increase the production of bio-oil and decrease the production of char. Increasing the density of the feedstock decreases the gap between two biomass particles. This allows uniform heat transfer rates between the particles and thus maintaining a uniform temperature across the entire medium. To achieve high yields from a reactor, the feed of the reactor is treated before it is pyrolyzed in the reactor. Some of the most common types of pretreatment techniques are ground, followed by drying of the feedstock and extrusion of biomass to form pellets.

4.1.1 Mechanical extrusion

Other steps of size reduction involve the extrusion of the biomass under very high pressure through a cylindrical mesh. The biomass is extruded to form small cylindrical pellets which are fed to the reactor. This process decreases the moisture content of the pellets and increases the energy density of the biomass. This is done by subjecting the biomass mixture to high temperatures of greater than 300°C. The heating and the mixing process result in the breaking of many cellulose crystals. However, the various steps required for the process require a lot of energy and therefore the entire process becomes costly [39].

4.1.2 Crushing and grinding

The main purpose of grinding the biomass is to achieve a smaller feed size. Decreasing the size of the feed increases the pyrolysis performance of the reactor. During the grinding process, the crystallinity of the cellulose is reduced. The grinding process involves various steps such as chipping, milling, or grinding. During the chipping process, the biomass is reduced to a size of 10–30 mm. This process increases the heat and mass transfer properties of the feed. However, the bio-oil yield from the product of the chipping process is very low. The yield of the reactor can be increased by reducing the size of the feed even more. The grinding/milling process can be used to reduce the feed size to 0.2 mm. This increases the density of the feed. During the grinding process, the cellulose crystallinity of the biomass is broken by the shear forces acting on it. The product from the milling process will have the highest heat transfer properties.

There are various types of milling processes, such as hammer milling, two-roll milling, colloid milling, and vibratory milling. These processes help in improving the digestibility of lignocellulosic materials. Hammer milling involves hammering the feed to produce finer bundles of feed. Two roll milling use the pressure produced between two cylindrical rollers for milling. The vibratory milling process is the most effective milling process, and involves vibrating balls to grind the biomass inside. The balls present inside the machine will be of different sizes and thus the feed product from the process will be very fine. Another commonly used milling process is disk milling, where the cellulose present inside is hydrolyzed to produce fibers. This process uses very little energy to run. These advantages make milling processes very effective in pretreatment methods for different biomass materials. The grinding process reduces the size of the feed by applying compressive forces from all directions. This increases the specific surface area of each particle and thus it becomes more exposed to the external hot surface during the pyrolysis reaction, which increases the yield of the reactor. However, the high cost of the equipment used for grinding will increase the overall cost of the operation.

4.2 Chemical techniques

Chemical pretreatment techniques use acids and alkalis to reduce the lignin content in lignocellulosic biomass. These techniques improve the properties of the pyrolysis products as they remove unwanted inorganic materials. Different techniques are used here. Sulfates, phosphates, carbonates, and chlorides are the impurities found in biomass, which affect the end-pyrolyzed products. Metal impurities found in biomass include sodium, potassium, calcium, and magnesium. NaOH, KOH, $Ca(OH)_2$, and NH_4OH are the alkaline solutions at low temperatures with high concentrations used. These alkaline solutions improve the biomass structure. They also partially remove the lignin, hemicellulose, and cellulose from biomass. NaOH and KOH are the most widely used of all alkaline solutions [3].

4.2.1 Dilute acid

The generation of a high number of inhibitory products such as furfurals, phenolic acids, 5-hydroxymethylfurfural, and aldehydes makes the acid treatment less appealing, although it is a widely used conventional pretreatment method for lignocellulosic feedstocks. Highly sustainable material is required for building the reactor due to the highly corrosive nature of the acids and the experimental conditions employed. Still, on an industrial scale, this method is widely used. Two types of acid pretreatments have been developed based on the type of end application:

- Low temperature (<120°C) for long duration (30–90 min),
- High temperature (above 180°C) for a short duration (1–5 min).

As the acid hydrolyzes the biomass into fermentable sugar, the enzymatic hydrolysis step is mostly avoided, however extensive washing is needed for the removal of acids before fermentation. Shrinking-bed, percolation, batch, flow-through, countercurrent reactors, and plug flow countercurrent reactors are some of the types of reactors developed. Before further processing, inhibitors generated in the process are removed. The process is economically feasible as the highly concentrated acid is recovered after hydrolysis. Some of the more commonly used of the many suitable acids are oxalic acid, maleic acid, and sulfuric acid.

4.2.2 Mild alkali

Mostly, alkali pretreatment methods are performed at room temperature and pressure. The hydroxyl derivatives of potassium, calcium, sodium, and ammonium salts are widely used as alkali reagents. Among these, the most effective was found to be sodium hydroxide. Totals of 60% lignin and 80% hemicellulose were extracted at the optimum condition when using 1.5% sodium hydroxide at 20°C for 144 h. On softwoods with a lignin content greater than 26%, no effect of dilute NaOH was observed. With alkali pretreatment, the solubility of cellulose and hemicellulose is very low as compared to acid pretreatment. After disrupting the lignin structure, increasing the internal surface area of cellulose, and decreasing the degree of polymerization and crystallinity, the solubility improves. In contrast to other pretreatment processes, the conditions for a mild alkali process are less harsh. Ambient conditions are sufficient for alkali pretreatment except if the pretreatment is needed to be carried out for a longer duration, when high temperature is required. In the end, to remove the inhibitors as well as lignin, a neutralizing step is required.

4.2.3 Ozonolysis

Ozonolysis is a pretreatment technique used for decreasing the lignin content in lignocellulosic biomass. It mainly reduces lignin and hardly affects cellulose and hemicellulose. Vidal and Molinier designed and developed a small-scale laboratory ozonolysis setup. This process has to be performed at ambient pressure and temperature. The process is eco-friendly as it does not produce any toxic inhibitors. It also does not affect the enzymatic hydrolysis and the yeast fermentation processes, which are the pre- and posttreatment processes, respectively. The affecting parameter is the moisture content of the specific biomass. The higher the moisture content, the lower the value of lignin oxidation. It is an expensive method, as a high amount of ozone is required to make it. This makes this process unlikely for an industrial scale.

4.3 Physiochemical techniques

Physiochemical techniques are the combination of both physical and chemical techniques, and one of the most important methods to obtain biomass pyrolytic products. This process decreases the thermal stability and makes changes to the components of biomass to obtain the destruction of lignocellulose. A variety of physiochemical treatments have been developed including acid and alkali pretreatments, hydrothermal pretreatment, ammonia fiber expansion, and steam explosion.

4.4 Steam explosion

Steam pretreatment is a common physiochemical technique used for the pretreatment of lignocellulosic biomass (Fig. 3.10). This process subjects the biomass to a high pressure of 0.7–4.8 MPa at elevated temperatures of between 160 and 260°C for a certain period of time. This causes hydrolysis and hemicellulose to be released. The steam enters the biomass in such a way that the walls of the fibers are expanded, which leads to partial hydrolysis and the accessibility of enzymes for cellulose is increased.

The first step is to load the biomass into a steam explosion vessel. Next is to add water into the vessel by checking the range of water to biomass ratios accordingly for the biomass component. Then, the vessel is heated to 160 and 260°C with pressure 0.7–4.8 MPa. The mixture inside is held for a specific period. This is to stimulate the hydrolysis of hemicellulose. This is followed by destruction of the mixture as the pressure is decreased. The treated biomass is collected through a discharge valve. The last step is to dry the resultant biomass at 105–110°C for 10–12 h [40].

This technique has been found to be effective for the pretreatment of hardwoods and agricultural residues. The acid catalyst is added for the softwood pretreatment technique. The advantages of this technique are as follows:

1. Limited use of chemicals
2. Low energy requirement
3. No recycling cost
4. Environment friendly.

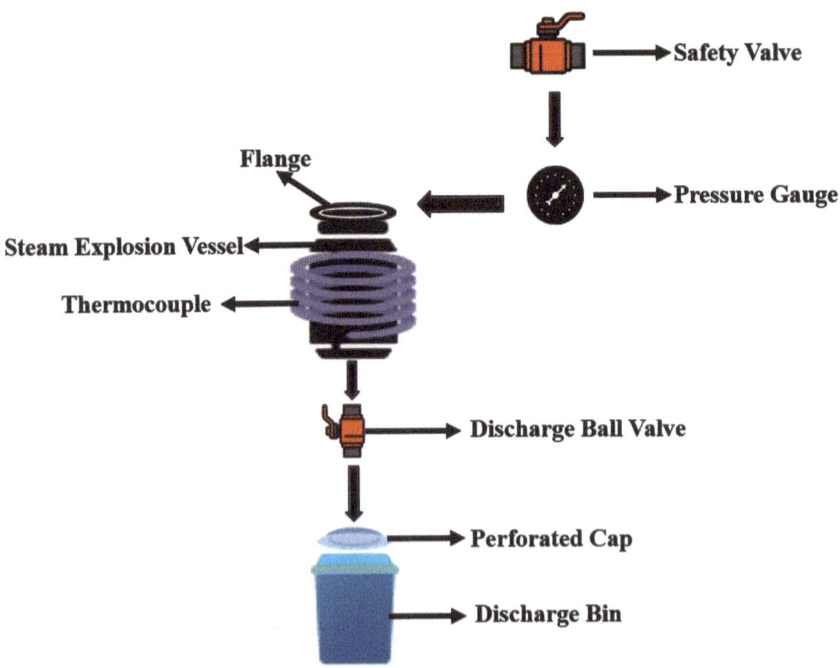

Figure 3.10 Steam explosion process.

4.4.1 Ammonia-based techniques

This process is carried out in a special reactor, as illustrated in Fig. 3.11. The reactor is equipped with a pressure and temperature controller. It has an inflow for liquid ammonia. A temperature of 60–120°C and pressure of 2 MPa are applied. This setup is maintained for 10–60 min in a closed vessel. The biomass content is placed in the reactor. The mixture of biomass and ammonia is heated. Then, the pressure is rapidly decreased by opening the vent valve. This rapid release results in the evaporation of ammonia, and the temperature also starts to decrease. This process removes the acetyl groups on hemicellulose, decrystallization of cellulose and lignin–carbohydrate complex linkages in lignin. Table 3.4 shows the differences between processes like AFEX, ARP, and SAA.

4.4.2 Wet oxidation

One of the simplest methods of biomass pretreatment techniques is wet oxidation. This process works by treating the biomass with oxygen with water or hydrogen peroxide at high temperatures (>120°C for 30 min). The efficacy of this method depends on three factors:

1. Temperature
2. Oxygen pressure
3. Reaction time.

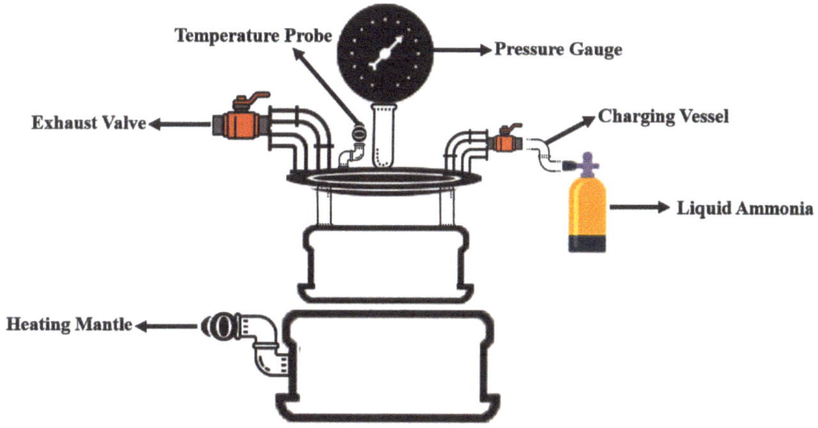

Figure 3.11 Ammonia fiber explosion reactor setup.

Table 3.4 Differences between AFEX, ARP, and SAA [39].

Ammonia fiber explosion (AFEX)	Ammonia recycle percolation (ARP)	Soaking aqueous ammonia (SAA)
High temperature 60–90°C	High temperature 60–90°C	Ambient temperature 30–60°C
Treats with liquid ammonia	Treats with liquid ammonia	Treats with aqueous ammonia
Pressure above 3 MPa	Pressure below 3 MPa	Pressure below 3 MPa

When the temperature is greater than 170°C, the water behaves like an acid and catalyzes the hydrolytic reactions taking place. Hemicelluloses are broken down into monomers which are small and in a pentagonal structure. Lignin undergoes oxidation. Cellulose is hardly affected by this method. Due to the high cost of hydrogen peroxide, this method is not used in the industrial scale treatment of biomass. Initially, this method was used for wastewater treatment and management and soil remediation.

4.4.3 Biological techniques

In comparison to physical and chemical pretreatments, biological methods are slower, but they consume less energy and leave a better environmental footprint. The microbial consortium is commonly used for the pretreatment of lignocellulosic biomass to improve biogas production. The cellulose and hemicellulose components are degraded by certain microbes chosen specifically from the natural environment [41]. Biological pretreatment of lignocellulose biomass is considered to be one of the most economical and eco-friendly treatment methods because it is carried out under ambient temperature and pressure and needs no energy requirements externally or chemical inputs. This makes the process more cost-effective as compared to its counterparts which involve physical and chemical methods. Moreover, they also manage to improve the biomass composition to produce high-value-added chemicals and energy-rich pyrolytic products. The result of the biological treatment of lignocellulose biomass is to degrade or depolymerize complex compounds, such as lignin, into their monomer units. Biological treatment of biomass can also decompose the main linkages between lignin and hemicellulose or lignin and cellulose, resulting in decreasing the activation energy and increasing the rate of reaction during thermochemical conversion of biomass at lower temperatures. Brown, white, and soft-rot fungi, which mainly degrade lignin and hemicellulose, are mainly used in these biological pretreatment techniques with a small amount of cellulose. Pretreatment of wheat straw by fungi (fungal isolate RCK-1) for 10 days increased fermentable sugars and resulted in a decrease in fermentation inhibitors. Even though the biological pretreatment is highly intriguing, the rate of hydrolysis of lignocellulosic fractions is too slow, which severely hampers it as a potential pretreatment method on an industrial scale. For more profitable results in biological pretreatment at a par with other pretreatment methods, more *Basidiomycetes* fungi should be tested for their ability to delignify the biomass effectively at a faster rate.

4.5 Thermal techniques

4.5.1 Densification

The major limitation of employing biomass as the feedstock for biofuel production is its low density. Low density means more volume, which makes it difficult to shelf, transport, and also could lead to very slow heat transfer during pyrolysis. A remedy for this limitation could be densification techniques which can increase the density by up to 10 times, making the biomass very suitable for pyrolysis. The compaction process of biomass by employing mechanical force to create uniform-sized solid pellets is usually referred to as "biomass densification." Compaction using a screw extruder is done in three steps. (1) Removing air from the voids which make the substance dense and in turn increases the surface area, which results in heat generation. (2) The heat generated from the previous step helps make the biomass soft, at this stage upon applying high-pressure results in the formation of local bridges and interlocking of particles. (3) Further compression done in a tapered die forms solid briquettes with uniform density.

4.5.2 Dry torrefaction

Dry torrefaction refers to the thermal pretreatment of biomass carried out at a temperature of 200−300°C in an inert atmosphere with low heating rates, which improves the biomass structure to produce better quality biofuels and thereby improving the economic feasibility of the pyrolysis process. The primary end-product of the torrefaction process is a solid product. In ideal conditions, three substances, hemicellulose (large portion), and cellulose, and lignin (very small portion), are thermally degraded to yield optimum qualities. Following the temperatures selected in the pretreatment, the dry torrefaction process can be categorized into three types: light at 200−235°C, mild at 235−275°C, and severe at 275−300°C. The parameters such as temperature, residence time, local gaseous environments (sweep gases), and pressure can play an influential role in the torrefaction process. This process improves many characteristics of biomass such as heating value, hydrophobicity, grindability, and particle size distribution upon size reduction. Table 3.5 illustrates differences between mild and severe torrefaction.

4.5.3 Microwave

Microwave pretreatment is one of the techniques used for the pretreatment of biomass. Various reasons for using this technique are as follows:

1. Low energy requirement
2. Easy operation
3. Minimum generation of inhibitors
4. High heating capacity in short duration.

Table 3.5 Differences between mild and severe torrefaction [40].

Mild torrefaction	**Severe torrefaction**
235−275°C	275−300°C
Depolymerizes hemicellulose and cellulose	Degrades hemicellulose, cellulose, and lignin
Torrefaction of cellulose—breaks glycosidic, hydrogen bonds	Torrefaction of lignin—breaks ether bonds
In cellulose structure, depolymerizes free hydroxyl groups on the glucose ring	Promotes demethoxylation reaction in a benzene ring structure

The addition of mild-alkali reagents is used for a more effective breakdown process. The higher microwave power leads to higher delignification and glucose production. An example is quoted for better understanding. NaOH and H_2SO_4 are made to react with microwaves, and are used for the pretreatment of *Miscanthus*. It resulted in a 12 times high sugar yield when compared to conventional heating during halftime. This was due to the predisruption of crystalline cellulose and the solubility of lignin in the chemicals.

4.5.4 Ultrasound

Ultrasonication is one of the techniques used for the pretreatment of biomass. Research scholars in laboratories have found that this process is a feasible pretreatment method. In this process, ultrasound waves produce both physical and chemical effects. This process leads to the formation of small bubbles with a cavity which helps in the rupture of cellulose and hemicellulose, which increases accessibility to the cellulose-degrading enzymes. This is the process of effective breakdown of cellulose into simpler reducing sugars. The influencing parameters of the ultrasonic field are frequency, duration, geometry and type of reactor, the solvent used, biomass characteristics, kinetics, and reactor configuration. The duration of this ultrasonication process has a maximum effect in the process. There is a certain time limit for sonication beyond which no additional effect in delignification is produced. The power of the sonication is determined by the frequency of sonication. The power of sonication is also an important factor affecting the lignocellulosic pretreatment. A 10–100 kHz ultrasound frequency is sufficient for the cell breakage process. High sonication adversely affects the pretreatment process. High power leads to bubble formation near the tip of the ultrasound transducer, which hinders the transfer of energy to the liquid medium. Table 3.6 illustrates the benefits of pretreated feedstocks.

Table 3.6 Benefits of pretreated feedstock for bio-oil production [37].

Feedstock pretreatment properties	Postpretreatment status	Economic benefits
Thermal degradation	Low acidic value	High profit
Grindability	Small uniform-sized particles for transferring heat in reactors	Low grinder operation cost
Homogeneity	Pyrolysis reactor has predictable conversion performance	Low management cost
Low moisture content	Low water level in bio-oil	Drying is not required
High heat value	Bio-oil has a high heat value	A combustion chamber is not required

5. Influence of reaction conditions on product

Generally, the pyrolysis reaction is carried out in an inert environment using carrier gases like nitrogen, helium, argon, etc. In certain cases, steam is also used as a carrier gas as it has shown augmentation of the bio-oil yield by reducing the secondary reactions. It is clear that the carrier gas plays a major role in determining the pyrolysis mechanism. Similarly, the pyrolysis reaction is also affected by several other factors such as temperature, pressure, heating rate, catalyst concentration, residence time, etc. Among these, parameters such as pyrolysis temperature, heating rate, and residence time greatly affect the product composition and yield. The following section gives a brief idea of these parameter effects on pyrolysis.

5.1 Temperature

The decomposition of biomass during pyrolysis takes place through different stages. Initially, reactions like evaporation, cleavage of bonds, free radical formation, formation of carboxyl, and carbonyl groups take place at a lower temperature range of 120–205°C. The second stage ranges between 200 and 600°C, where major products such as bio-oil, syngas, and char are produced. Finally, above 600°C, the char decomposition takes place at a slow rate which generates high-carbon solid residues. From these stages, it becomes evident that the product composition mainly depends on the pyrolysis temperature. The higher temperature in the third stage supports dehydration and decarboxylation reactions, which directly augment aromatic, polar, and aliphatic compounds in the products. This also affects the yield of gaseous product along with its composition. Several research works have been reported in studying the effect of pyrolysis temperature on product yield and quality [42,43]. Conversely, at lower temperature, water yield is increased, whereas the phenolic components were decreased in the bio-oil due to insufficient energy to break the lignin structure of the biomass. The water production is observed to be ceased at around 580°C in most of the biomass pyrolysis reactions. Also, at this temperature range of around 400–500°C, the peak bio-oil yield has been observed for biomass pyrolysis. It is worth noting that low-temperature pyrolysis or slow pyrolysis favors char formation due to a lack of energy for bond breakage and curbing of secondary reactions. Meanwhile, higher temperature pyrolysis supports the disruption of C–H, C–C, and C–O bonds, which catalyzes the secondary reaction and leads to augmentation of the gaseous product yield. It was also evident that an effective change in pyrolysis temperature can control the carbon content and pH value of the char produced. On the other hand, temperature changes influence directly the variation of salient functional groups in the products. Some of the key result findings are given in Table 3.7.

5.2 Heating rate

Next to the temperature, the heating rate plays a vital role in determining the product yield. In general, the overall pyrolysis process can be divided into two major types

Table 3.7 Key findings on the effect of temperature.

Feedstock	Temperature range (in °C)	Inference	References
Calophyllum inophyllum seed cake	450–550	Highest bio-oil yield is obtained at 500°C, above which the gas yield increased	[42]
Straw and lignosulfonate	200–600	Char yield dropped along with an increase in carbon, pH, and stability with temperature rise	[44]
Safflower seed cake	400–600	Char obtained at 600°C can be directly used in energy applications due to superior properties	[45]
Coconut fiber	350–600	Char yield decreased along with an increase in pH, ash, surface area, and porosity with an increase in temperature	[46]
Rapeseed	400–700	Both chemical and surface properties are significantly affected by temperature	[47]
Wheat straw	475–575	Maximum bio-oil yield about 60 wt.% is observed around 525–550°C and the char decreased at high temperatures	[48]

based on heating rate, as slow pyrolysis (up to 100°C/min) and fast pyrolysis (above 1000°C/min). The heating rate augmentation catalyzes the effective depolymerization of elemental units such as cellulose and hemicellulose. The elevated heating rate reduces the residence time of short-chain volatile organic compounds and curbs the secondary reactions. Also, it promotes the cracking reactions and can produce more liquid product (bio-oil), along with less solid product (char). Meanwhile, lower heating rates augment the intrahydrogen bonds of the cellulose fraction leading to the initiation of dehydration reactions. The heating rate variation not only affects the product yield but also influences the composition and properties of pyrolysis products. The higher heating rates stabilize the bio-oil formation with a lower amount of water due to a reduction of secondary cracking of hydrogen bonds. The oxides of carbon are increased in gas-phase products due to elevated heating rates. It can also be observed that char with a smaller pore size can be generated with high heating rate experiments. Table 3.8 displays the key findings regarding the heating rate effects on pyrolysis products.

5.3 Residence time

The residence time of vapor and solid components inside the reactor significantly affects the yield and properties of the products. The shorter vapor residence time limits

Table 3.8 Key findings on the effect of heating rate.

Feedstock	Heating rate range (in °C/min)	Inference	References
Waste wood	5−350	At high heating rates, gas yield increases along with a decrease in liquid yield. The properties of products are affected by the heating rate	[49]
Waste tires	60−6000	High heating rate favored the cracking reaction and formation of radical fragments to form methane, hydrogen sulfide, and hydrogen	[50]
Nylon	5-40	The degradation temperature of nylon is shifted to the higher side with augmentation of the heating rate	[51]
Mixed biomass	60−3000	The bio-oil yield is observed to be constant after a heating rate of 600°C/min	[52]
Sugarcane bagasse	5−1200	With an increase in heating rate, the products of dehydration and fragmentation are increased	[53]

the secondary cracking of volatiles, thereby increasing the bio-oil yield with a reduction of char yield. On the other hand, an elevated residence time ensures complete decomposition of the feedstock, thereby augmenting the gas yield. Therefore, it becomes mandatory to fit a suitable residence time to obtain the yield of desired product. The longer residence time leads to secondary cracking of volatiles and generates secondary products. In addition, the gas yield will significantly increase owing to the high rate of secondary reactions.Table 3.9 depicts some of the key research findings with respect to residence time effects on pyrolysis products.

6. Advancements in biomass pyrolysis

In recent decades, the thermochemical conversion process has achieved new heights in terms of research and monetary aspects. Researchers around the globe have focused to simplify the methodology in order to make it simple and affordable. Meanwhile, certain research activities have been carried out to upgrade the existing technology by optimizing the parameters. Apart from these, new technologies have also been developed for gaining more benefits from the available resources. This section is devoted to discussing such advancements in the field of pyrolysis in recent decades.

The conventional pyrolysis process has been proven to be efficient for a wide range of biomass feedstocks. The pyrolysis conditions can be altered to obtain the desired products, as discussed in earlier sections. Meanwhile, altering the method of heating

Table 3.9 Key findings on the effect of residence time.

Feedstock	Residence time	Inference	References
Waste wood	32 and 64 min	The high temperature and longer reaction time during the pyrolysis of waste wood in screw reactor favor char yield	[54]
Sewage sludge and cotton stalks	30–150 min	The surface area char is observed to be increased with a residence time of 30–90 min, above which the value is decreased	[55]
Wheat straw pellets	100–200 s	The thermodynamic irreversibility of the pyrolysis system is considerably reduced while operated at lower residence times	[56]
Plastic	0.2–2 s	With a lower volatile residence time of 0.2 s, the soot emission is not observed till reaching a temperature of 1200°C	[57]
Rapeseed stem	10–100 min	The surface area and morphology of char are significantly influenced by the residence time	[58]

has a great impact on product distribution and product quality. Microwave pyrolysis is one such advanced pyrolysis process which employs microwaves to heat the biomass sample. The heat is generated through friction between molecules due to their motion caused by the microwaves. The higher heating rate, volumetric core heating, and rapid process make microwave pyrolysis as superior method as compared to the conventional one. On considering the yield, along with common parameters, certain specific parameters like microwave reactor design, heating manner, and microwave power level also play important roles. In spite the variations in yield of products observed in various literature works, some common facts are also recorded. The majority of researchers observed high solid and gases product yields, along with lower liquid yields in microwave pyrolysis, which is one of the major challenges in employing microwave pyrolysis for liquid fuel production. The char obtained through microwave heating is observed to be more porous and possessed a high surface area. Meanwhile, owing to the higher hydrogen content, the calorific value of the gaseous product is fairly high. The overall performance of the microwave pyrolysis can still be improved by using catalysts. It is worth noting that the preexponential factor and activation energy are always less for microwave pyrolysis as compared to those of a conventional process. On the whole, the microwave pyrolysis can be considered to be a suitable alternative to the traditional pyrolysis process in energy extraction from biomass.

Copyrolysis and tripyrolysis are emerging technologies in the field of thermochemical conversion of biomass. The combination of two different feedstocks with synergistic effects, has been proved to enhance the quality and yield of the products. The organic biomass component is blended with an inorganic feedstock (generally plastics) to enhance the high-temperature pyrolysis characteristics. This synergistic performance considerably reduces the unwanted aldehydes, ketones, and sugars, along with augmentation in alcohols in the fuel product [58]. Gou et al. conducted a kinetic study on the tripyrolysis of corn stalk, polyethylene, and anthracite. The authors elucidated the effects of different ratios and temperatures on the product distribution and kinetic models [59].

7. Conclusion

As the world is moving toward greater industrialization, the accumulation of waste becomes inevitable. With this in mind, technologies have been evolving to manage the wastes and turn them into useful products. Thermochemical conversion is one such prominent process to convert wastes into useful commercial resources. This chapter has outlined the salient features of the pyrolysis process and provided a wide insight to readers about the reactors, effects of operating conditions, and advancements in this area. The different pretreatment techniques, along with their merits and demerits, have also been discussed to provide a deep understanding to readers. On the whole, this chapter provides preliminary knowledge for the budding engineers who are starting their career in the field of pyrolysis.

References

[1] T. Donohue, Welcome to biofuels, Biofuels 1 (1) (2010) 1–2.
[2] M.R.L.V. Leal, A. Walter, The sustainability of biofuels in view of the EU renewable energy directive, Biofuels 1 (6) (2010) 799–803.
[3] Z. Echresh Zadeh, et al., Recent insights into lignocellulosic biomass pyrolysis: a critical review on pretreatment, characterization, and products upgrading, Processes 8 (2020) 799.
[4] S. Yılmaz, H. Selim, A review on the methods for biomass to energy conversion systems design, Renewable and Sustainable Energy Reviews 25 (2013) 420–430.
[5] M. Sameeroddin, et al., Renewable energy: fuel from biomass, production of ethanol from various sustainable sources by fermentation process, Materials Today Proceedings (2021).
[6] M. Shahabuddin, et al., A review on the production of renewable aviation fuels from the gasification of biomass and residual wastes, Bioresource Technology 312 (2020) 123596.
[7] D.C. Elliott, et al., Hydrothermal liquefaction of biomass: developments from batch to continuous process, Bioresource Technology 178 (2015) 147–156.
[8] M.I. Jahirul, et al., Biofuels production through biomass pyrolysis—a technological review, Energies 5 (12) (2012).
[9] A.G.W. Bradbury, Y. Sakai, F. Shafizadeh, A kinetic model for pyrolysis of cellulose, Journal of Applied Polymer Science 23 (11) (1979) 3271–3280.
[10] B. Hu, et al., Catalytic mechanism of sulfuric acid in cellulose pyrolysis: a combined experimental and computational investigation, Journal of Analytical and Applied Pyrolysis 134 (2018) 183–194.

[11] M.J. Antal Jr., G. Varhegyi, Cellulose pyrolysis kinetics: the current state of knowledge, Industrial & Engineering Chemistry Research 34 (3) (1995) 703–717.
[12] A.L. Brown, D.C. Dayton, J.W. Daily, A study of cellulose pyrolysis chemistry and global kinetics at high heating rates, Energy & Fuels 15 (5) (2001) 1286–1294.
[13] M. Wu, G. Várhegyi, Q. Zha, Kinetics of cellulose pyrolysis after a pressurized heat treatment, Thermochimica Acta 496 (1) (2009) 59–65.
[14] Y.-C. Lin, et al., Kinetics and mechanism of cellulose pyrolysis, Journal of Physical Chemistry C 113 (46) (2009) 20097–20107.
[15] J.B. Wooten, J.I. Seeman, M.R. Hajaligol, Observation and characterization of cellulose pyrolysis intermediates by 13C CPMAS NMR. A new mechanistic model, Energy & Fuels 18 (1) (2004) 1–15.
[16] P.R. Patwardhan, et al., Distinguishing primary and secondary reactions of cellulose pyrolysis, Bioresource Technology 102 (8) (2011) 5265–5269.
[17] X. Zhou, et al., A critical review on hemicellulose pyrolysis, Energy Technology 5 (1) (2017) 52–79.
[18] R.S. Miller, J. Bellan, A generalized biomass pyrolysis model based on superimposed cellulose, hemicelluloseand liqnin kinetics, Combustion Science and Technology 126 (1–6) (1997) 97–137.
[19] Y. Peng, S. Wu, The structural and thermal characteristics of wheat straw hemicellulose, Journal of Analytical and Applied Pyrolysis 88 (2) (2010) 134–139.
[20] A.V. Bridgwater, D. Meier, D. Radlein, An overview of fast pyrolysis of biomass, Organic Geochemistry 30 (12) (1999) 1479–1493.
[21] E. Adler, Lignin chemistry—past, present and future, Wood Science and Technology 11 (3) (1977) 169–218.
[22] V.B.F. Custodis, et al., Mechanism of fast pyrolysis of lignin: studying model compounds, The Journal of Physical Chemistry B 118 (29) (2014) 8524–8531.
[23] X. Hu, M. Gholizadeh, Biomass pyrolysis: a review of the process development and challenges from initial researches up to the commercialisation stage, Journal of Energy Chemistry 39 (2019) 109–143.
[24] N. Awan, et al., Effect of operating parameters on production of bio-oil from fast pyrolysis of maize stalk in bubbling fluidized bed reactor, Polish Journal of Chemical Technology 18 (2016).
[25] F. Campuzano, R.C. Brown, J.D. Martínez, Auger reactors for pyrolysis of biomass and wastes, Renewable and Sustainable Energy Reviews 102 (2019) 372–409.
[26] Y. Yu, et al., Pyrolysis of rice husk and corn stalk in auger reactor. 1. Characterization of char and gas at various temperatures, Energy & Fuels 30 (12) (2016) 10568–10574.
[27] M. Mathew, L. Muruganandam, Pyrolysis of agricultural biomass using an auger reactor: a parametric optimization, International Journal of Chemical Reactor Engineering 15 (2017).
[28] Z. Xiong, et al., Evolution of coke structures during the pyrolysis of bio-oil at various temperatures and heating rates, Journal of Analytical and Applied Pyrolysis 134 (2018).
[29] M.A. Aziz, M.A. Rahman, H. Molla, Design, fabrication and performance test of a fixed bed batch type pyrolysis plant with scrap tire in Bangladesh, Journal of Radiation Research and Applied Sciences 11 (4) (2018) 311–316.
[30] A. Trendewicz, et al., One dimensional steady-state circulating fluidized-bed reactor model for biomass fast pyrolysis, Fuel 133 (2014) 253–262.
[31] B.M. Wagenaar, W. Prins, W.P.M. van Swaaij, Pyrolysis of biomass in the rotating cone reactor: modelling and experimental justification, Chemical Engineering Science 49 (24, Part 2) (1994) 5109–5126.

[32] G. Luo, et al., Pyrolysis of whole wood chips and rods in a novel ablative reactor, Fuel 194 (2017) 229–238.
[33] B. Gómez-Monedero, et al., Pyrolysis of red Eucalyptus, camelina straw, and wheat straw in an ablative reactor, Energy & Fuels 29 (3) (2015) 1766–1775.
[34] M.D.M. Hasan, et al., Grinding pyrolysis of Mallee wood: effects of pyrolysis conditions on the yields of bio-oil and biochar, Fuel Processing Technology 167 (2017) 215–220.
[35] R.W. Nachenius, et al., Chapter two—biomass pyrolysis, in: D.Y. Murzin (Ed.), Advances in Chemical Engineering, Academic Press, 2013, pp. 75–139.
[36] N.S. Ab Rasid, et al., Recent advances in green pre-treatment methods of lignocellulosic biomass for enhanced biofuel production, Journal of Cleaner Production 321 (2021) 129038.
[37] H. Rezaei, et al., Woody Feedstock Pretreatments to Enhance Pyrolysis Bio-oil Quality and Produce Transportation Fuel, 2018.
[38] R.D. Kasparbauer, The effects of biomass pretreatments on the products of fast pyrolysis, in: Mechanical Engineering, Iowa State University, 2009, p. 307.
[39] A.K. Kumar, S. Sharma, Recent updates on different methods of pretreatment of lignocellulosic feedstocks: a review, Bioresources and Bioprocessing 4 (1) (2017) 7.
[40] R. Kumar, et al., Lignocellulose biomass pyrolysis for bio-oil production: a review of biomass pre-treatment methods for production of drop-in fuels, Renewable and Sustainable Energy Reviews 123 (2020) 109763.
[41] T. Kan, V. Strezov, T.J. Evans, Lignocellulosic biomass pyrolysis: a review of product properties and effects of pyrolysis parameters, Renewable and Sustainable Energy Reviews 57 (2016) 1126–1140.
[42] R. Sakthivel, K. Ramesh, Influence of temperature on yield, composition and properties of the sub-fractions derived from slow pyrolysis of *Calophyllum inophyllum* de-oiled cake, Journal of Analytical and Applied Pyrolysis 127 (2017) 159–169.
[43] W. Hu, et al., Combustion behaviors of molded bamboo charcoal: influence of pyrolysis temperatures, Energy 226 (2021) 120253.
[44] J. Zhang, J. Liu, R. Liu, Effects of pyrolysis temperature and heating time on biochar obtained from the pyrolysis of straw and lignosulfonate, Bioresource Technology 176 (2015) 288–291.
[45] S.A. Dhar, T.U. Sakib, L.N. Hilary, Effects of pyrolysis temperature on production and physicochemical characterization of biochar derived from coconut fiber biomass through slow pyrolysis process, Biomass Conversion and Biorefinery (2020).
[46] D. Angin, S. Sensöz, Effect of pyrolysis temperature on chemical and surface properties of biochar of rapeseed (*Brassica napus* L.), International Journal of Phytoremediation 16 (7–12) (2014) 684–693.
[47] N. Ibrahim, et al., Influence of reaction temperature and water content on wheat straw pyrolysis, World Academy of Science, Engineering and Technology, International Journal of Chemical, Molecular, Nuclear, Materials and Metallurgical Engineering 6 (2012) 919–925.
[48] C.E. Efika, J.A. Onwudili, P.T. Williams, Influence of heating rates on the products of high-temperature pyrolysis of waste wood pellets and biomass model compounds, Waste Management 76 (2018) 497–506.
[49] H. Wang, et al., Effect of high heating rates on products distribution and sulfur transformation during the pyrolysis of waste tires, Waste Management 118 (2020) 9–17.
[50] A.M. Pannase, et al., Decomposition of polyamide via slow pyrolysis: effect of heating rate and operating temperature on product yield and composition, Journal of Analytical and Applied Pyrolysis 151 (2020) 104886.

[51] J.L. Klinger, et al., Effect of biomass type, heating rate, and sample size on microwave-enhanced fast pyrolysis product yields and qualities, Applied Energy 228 (2018) 535–545.
[52] J. Montoya, et al., Effect of temperature and heating rate on product distribution from the pyrolysis of sugarcane bagasse in a hot plate reactor, Journal of Analytical and Applied Pyrolysis 123 (2017) 347–363.
[53] J. Solar, et al., Influence of temperature and residence time in the pyrolysis of woody biomass waste in a continuous screw reactor, Biomass and Bioenergy 95 (2016) 416–423.
[54] Z. Wang, et al., Effects of residence time on characteristics of biochars prepared via co-pyrolysis of sewage sludge and cotton stalks, Journal of Analytical and Applied Pyrolysis 142 (2019) 104659.
[55] G. Greco, et al., Effects of slow-pyrolysis conditions on the products yields and properties and on exergy efficiency: a comprehensive assessment for wheat straw, Applied Energy 279 (2020) 115842.
[56] X. Wang, et al., Soot formation during polyurethane (PU) plastic pyrolysis: the effects of temperature and volatile residence time, Energy Conversion and Management 164 (2018) 353–362.
[57] B. Zhao, et al., Effect of pyrolysis temperature, heating rate, and residence time on rapeseed stem derived biochar, Journal of Cleaner Production 174 (2018) 977–987.
[58] J. Yang, et al., Fast co-pyrolysis of low density polyethylene and biomass residue for oil production, Energy Conversion and Management 120 (2016) 422–429.
[59] X. Gou, et al., Tri-pyrolysis: a thermo-kinetic characterisation of polyethylene, cornstalk, and anthracite coal using TGA-FTIR analysis, Fuel 252 (2019) 393–402.

Effects of an hourly tariff on the electricity power sector—the Honduran model

4

Bryan R. Jovel-Zelaya [1], *Gerson O. Vargas-Godoy* [1], *Wilfredo C. Flores* [1], *Harold R. Chamorro* [2], *Jacobo Aguillon-Garcia* [3], *Vijay K. Sood* [4], *Ameena Al-Sumaiti* [5], *Paul Vasquez-Miranda* [6], *Francisco Gonzalez-Longatt* [7] and *Charalambos Konstantinou* [8]

[1]Universidad Tecnológica Centroamericana, Tegucigalpa, Honduras; [2]KTH, Royal Institute of Technology, Stockholm, Sweden; [3]Korea Advanced Institute of Science and Technology, Daejeon, Korea; [4]University of Ontario, Institute of Technology, Oshawa, ON, Canada; [5]Advanced Power and Energy Center, Khalifa University, Abu Dhabi, United Arab Emirates; [6]Escuela Politécnica Nacional, Universidad Politécnica Salesiana, Quito, Ecuador; [7]University of South-Eastern Norway, Porsgrunn, Norway; [8]Division of Computer, Electrical and Mathematical Sciences and Engineering, King Abdullah University of Science and Technology (KAUST), Thuwal, Saudi Arabia

1. Introduction

The grid in the electrical system in Honduras is undergoing a transformation in each of its components, from flexible generation systems, transmission, and distribution lines with greater energy transport capacity, and dynamic demand-side management adaptable to the requirements for each type of user. The flexibility of the grid is defined as its ability to adapt to variable conditions without losing the balance between power generation and its consumption. By increasing the penetration of distributed generation systems, including renewables (solar, wind, geothermal, etc.) in the national interconnected system, new technologies are required to fully link the behavior of the end user with the electrical grid. For this purpose, it is planned to have new transmission lines from the scale powers, new substations that allow transmission links, intelligent measurement at the system's borders (transformers from high-voltage to medium-voltage substations), and remote-controlled measurement equipment in the distribution network of the country.

To have flexibility in the network, specialized measurement equipment of hourly aggregated energy consumption is needed to know and generate consumption and power demand curves per hour. These types of equipment can be smart meters for medium- and low-voltage restorers, among others; in the same way, the generation of energy per hour influences the collection of the electricity rate (this must be included in the liberation of the electricity market in Honduras) and allows an hourly rate to be

Green Energy Systems. https://doi.org/10.1016/B978-0-323-95108-1.00004-5

charged to the end user. This equipment is telemanaged and programmed in time blocks.

The national predispatch is carried out on an hourly basis a day ahead of the physical dispatch of the units, using as a basis the hourly demand expected in each node of the power network, the generation units, and transmission facilities that are declared as available. The results of the national predispatch will include the nodal prices in the transmission system, the hourly productions of each power generation plant at each node, and the required complementary services.

The hourly rate becomes an essential requirement for the flexibility of the power network, with participation in the entire national interconnected system, from the expected hourly production of the generating units through renewable resources, hourly projections of demand per node of the transmission system, and the characteristic curve of the daily hourly demand in the network. In compliance with the aforementioned, new technologies are required that measure by time blocks, establishing effective payment to generators according to their dispatch time and connection node, and a charge based on generation and transport according to the time block demanded by the final consumer and, as shown in the analysis of this report, an economic saving reflected at the end user.

2. Current challenge in Honduras

Currently, there is no multihour tariff applied to the Honduran electricity sector, in comparison to other countries with larger scales of industry and commerce which apply these required tariffs for the reduction and implementation of efficient systems. However, there is a national law of hourly tariffs with a typical demand curve for multihour tariffs that would give a more efficient insight into the consumption of the general population of Honduras; this would economically help all sectors, since the demand curve would move at different times depending on the consumption of each sector.

3. Theoretical framework

3.1 Time tariff

Current tariffs in Honduras are structured in such a way that energy is used efficiently for the commercial, industrial, and residential subsectors. The hourly tariff is a form of electric tariff, where a price is established for the kilowatt hour (kWh) of energy consumption, which is a type of tariff that allows for savings on the energy bill.

Time discernment is a type of energy tariff that accounts for electricity consumption in two periods (day and night), with night hours being the cheapest, also known as off-peak hours. It is a highly endorsed tariff modality for those who concentrate much of their consumption at night. Not all consumers are interested in this tariff, since it usually implies a change in consumption habits.

In Honduras, the Electric Power Regulatory Commission[1] issued a statement establishing the parameters and conditions of users, where it is written that users must be charged through an hourly tariff system since June 2016. The challenge associated with this provision lies with the distribution by the National Electric Energy Company (ENEE, by its Spanish acronym), which must adapt its commercial system to comply with the requirements. In addition to this, it must jump to a change from one electricity consumption measurement technology to smart technology (smart meters) that can be configured or programmed for multiple scenarios for the hourly tariff.

To make a distribution time more in accord with reality, cut-off levels have been established on the demand profile that result in the same number of hours in each period; it is currently divided as shown in Table 4.1.

Table 4.1 Percentage of demand according to day classification.

	Peak	Semivalley	Valley
Working day	>93%	67%–93%	<67%
Saturday		68%–93%	<68%
Sunday		72.8%–93%	<72.8%

Source: AETS en asosiación con BCEOM y EDE Ingenieros, Tarifas multihorarias, socializacion e impacto - GAUREE II. EuropeAID, Tegucigalpa M.D.C., 2011.

3.1.1 General law of the electricity industry for hourly tariffs

For the analysis of the project, the Regulation for the Calculation of Provisional Tariffs, published by the Official Gazette (La Gaceta in Spanish) No. 34,012 of April 20 [3], was taken as a base. It is emphasized in the following articles.

Article 2. For the purposes of these regulations, the tariff for a particular user is a price vector that is applied to a corresponding consumption vector. The measurement shall record the consumption tariff components separately. The National Enterprise of Electric Energy (ENEE in Spanish) will calculate for each category of users a price vector with five components.

Article 3. For the purposes of calculating tariffs, CREE defines the following time blocks (see Table 4.2).

Article 4. Once the five price rates for each category of users have been determined, the ENEE will calculate in the case of small consumptions served in low voltage—particularly small residential consumption—simpler tariffs. For those small consumptions, the tariffs will consist of a fixed charge and a single energy price. The latter will be calculated as a monetary price that includes the price of power and the three prices of energy per hour block.

[1] CREE, by its Spanish acronym.

Table 4.2 Percentage of demand according to days classification.

Day type	Peak period		Middle period		Valley period	
	Total hours	Schedule	Total hours	Schedule	Total hours	Schedule
Working day	10	9–16 18–21	9	5–9 16–18 21–24	5	0–5
Saturday	2	11–12 19–20	16	6–11 12–19 20–24	6	0–6
Sunday or holidays	0	–	8	11–13 17–23	16	0–11 13–17 23–24

Source: Resolución CREE 148, Comisión Reguladora de Energía Eléctrica. Obtenido de Comisión Reguladora de Energía Eléctrica, 24 de Junio de 2019: https://cree.gob.hn/wp-content/uploads/2019/02/Reglamento-de-Tariffs.pdf.

Article 5. To calculate tariffs, the ENEE will study the demands of users, who will be grouped into five categories:

a. Residential
b. General low-voltage service (shops, offices, workshops, connected in low voltage)
c. Street lighting
d. Medium-voltage service
e. High-voltage service.

Article 6. The classification of users into categories is based on the consumption patterns of these groups and the level of electrical tension in which they are served. Categories that respond to other classification criteria such as "government" or "municipalities" will not be used. An installation of an industrial nature owned by the government, served in medium voltage, shall be classified in the category "medium-voltage service"; nevertheless, some government offices served in low-voltage service shall be classified in the category "general service in low voltage."

For the sake of the article, the information previously mentioned regarding the hourly blocks established regulations by CREE in 2016 will be used, the energy price of the first quarter of 2016 for each user in the "low-voltage" category will also be used as a reference. Medium- and high-voltage categories will be analyzed by hourly blocks. Industrial, commercial, and residential users who meet the requirements to establish a charge for an hourly tariff will be determined. Residential users who do not consume more than 500 kWh will not comply with the low-voltage hourly tariff. These end-users will be subject to a previously established fixed tariff. The users that will apply the tariff in medium voltage are those who are connected at a voltage of

13.8 kV or 34.5 kV and have the measurement directly connected at that voltage level. The users for which the high-voltage tariff applies are those connected at a voltage of 69, 138, or 230 kV (measured directly at that voltage level).

3.2 User's valid tariff and provisional tariff regulation

The current structure of the hourly tariff is updated quarterly and will be charged in the same way in each of the four periods of the year, with restrictions on days to total invoices (the bill arrives after between 28 and 32 days), according to the level of tension (low tension, medium tension, high tension) and consumption sector (residential, commercial, or industrial); fixed charges will be handled without time differentiation. These values are used to obtain a distinction of cost over time (hourly tariff). In 2016, CREE proposed in its quarterly fixed tariff to deliver a variant of the hourly tariff, where it adds to the current regulations, schedules that were analyzed in the power demand curve, obtaining costs to be charged for the consumption of electrical energy.

The correlation methodology to be used in this analysis starts from this information, first, obtaining a present value of the tariffs proposed in 2016–2020. Tariff sectors chosen for this analysis are the residential, commercial, and industrial sectors that represent most of the national electricity consumption (approximately 89%) [4,7].

There is not an analysis of the total consumption or users among these three sectors proposed due to the limitations explained in the following section.

4. Used methodology

This research is carried out based on the implementation of hourly tariffs to benefit users and the system by flattening the electrical demand curve of the Honduran electrical system. This is the result of research and analysis of electrical demand curves, types of consumers, strata, and possible application scenarios, among other aspects that make up the electrical subsector. A correlational study is used to determine if two variables are harmonized or not. This means analyzing whether an increase or decrease in one variable coincides with an increase or decrease in the other variable (Fig. 4.1).

The proposed methodology to address this issue has been used in different countries that apply the hourly tariff system. They analyzed their electrical system and each of the parameters that compose it to obtain a satisfactory result for the needs and problems that arise from having a fixed electric tariff. The goal of applying the correlation methodology is to compare the results obtained with the implementation of the hourly proposed tariff and the current fixed tariff, analyze the behavior of the electricity sector, and decide in which possible scenarios it can be applied, and the advantages that will be observed in the subsectors.

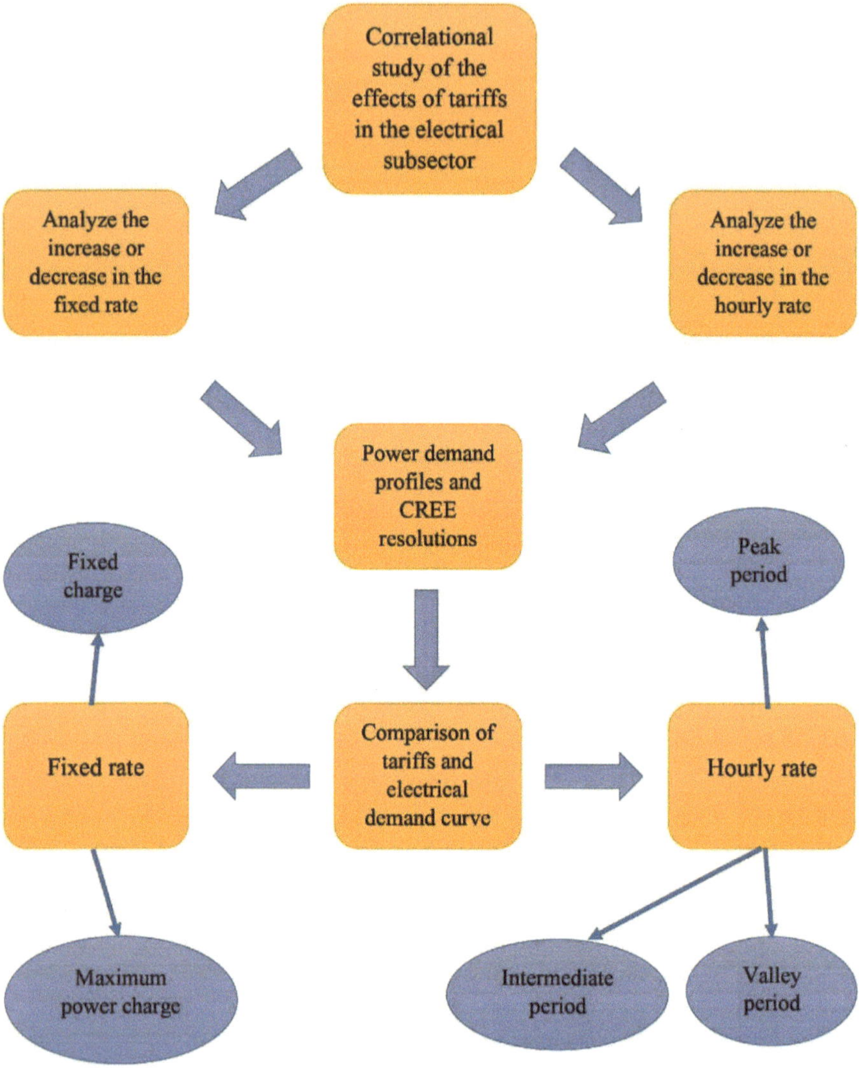

Figure 4.1 Flow diagram methodology to be used in the correlational study.

5. Analysis and results

5.1 Analysis of the curve of electricity consumption and power demand

Starting from the hourly tariff, the curve of electrical consumption and power demand are analyzed in such a way as to validate the sectors with the greatest contribution to them. In an established way, the residential, commercial, and industrial sectors are chosen for punctual analysis of users to which their load profiles are accessed to implement the correlation methodology with the current tariffs and proposed hourly tariff.

Table 4.3 Record of energy tariffs.

Resolution date	Residential (less than 50 kWh) (L)	Residential (more than 50 kWh) (L)	Low power (L)	Medium power (L)	High power (L)
May 2016	1.4558	3.6051	3.937	2.3989	2.2441
July 2016	1.4588	3.6051	3.937	2.3989	2.2441
November 2016	1.4911	3.6848	4.0315	2.4603	2.2961
March 2017	1.6217	4.0073	4.3197	2.7159	2.5487
May 2017	1.6214	4.0065	4.3114	2.7068	2.5401
August 2017	1.6137	3.9871	4.2884	2.6914	2.5269
December 2017	1.621	4.0051	4.314	2.7133	2.5472
March 2018	1.6427	4.0588	4.314	2.7133	2.5472
June 2018	1.6776	4.145	4.314	2.7299	2.5472
September 2018	1.8889	4.6671	4.7928	3.1611	2.9417
December 2018	3.643	4.7404	4.7373	3.0883	2.8755
March 2019	4.0274	5.2406	5.3266	3.4056	3.171
July 2019	3.9728	5.1696	5.2364	3.4006	3.1601
October 2019	3.8678	5.033	5.1195	3.2387	2.9952
December 2019	4.0088	5.2164	5.1945	3.3617	3.1463

Source: Resolución CREE 148, Comisión Reguladora de Energía Eléctrica. Obtenido de Comisión Reguladora de Energía Eléctrica, 24 de Junio de 2019: https://cree.gob.hn/wp-content/uploads/2019/02/Reglamento-de-Tariffs.pdf.

Table 4.3 presents the history of energy tariffs since 2016; it can be observed that the price for residential electricity, as the fluctuation of the price of residential consumption, has fluctuated given the various groups (housing less than 50 kWh, greater than 50 kWh, low power, medium power, and high power) (Table 4.3).

The tariff figure for energy gives us a representation of the price per kWh for the last 4 years, from May 2016 to December 2019 (Fig. 4.2).

The small consumptions served in low tension, particularly in the small residential consumptions, and the tariffs will consist of the fixed charge and a single energy price.

5.2 Update of energy costs by time tariff

For the purposes of calculating the tariffs, CREE defines the time blocks as shown in Table 4.4.

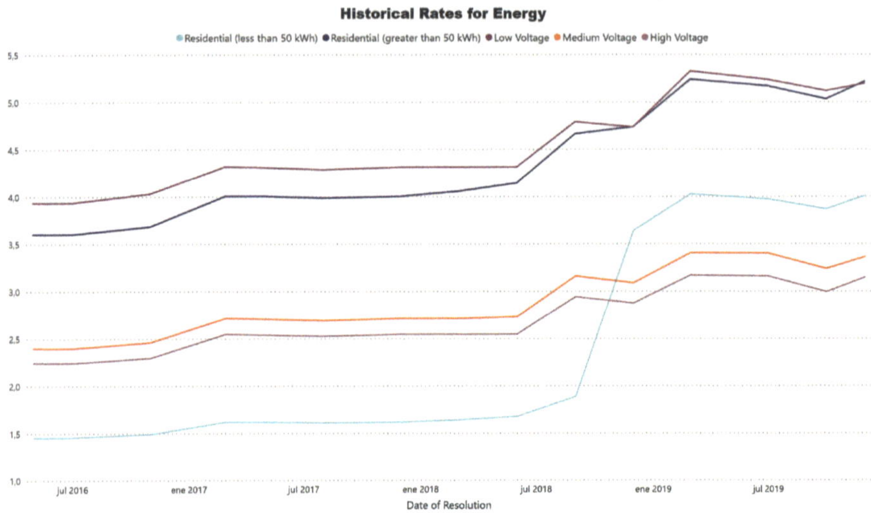

Figure 4.2 Historic tariffs by energy.
Source: Historial de Tariffs CREE, Comisión Reguladora de Energía Eléctrica. Obtenido de HISTORIAL DE TARIFFS CREE, Diciembre de 2019: https://www.cree.gob.hn/historial-de-Tariffs/; Observatorio de la Energía UNITEC HN, Observatorio de la Energía UNITEC HN. Obtenido de Gracia Pineda, 04 de Agosto de 2019: https://public.tableau.com/profile/gp17447#!/vizhome/EstadsticasElectricidadHonduras/EstadsticasENEE.

Table 4.4 Record of energy tariffs.

	Peak period		Middle period		Valley period	
Day type	**Total hours**	**Schedule**	**Total hours**	**Schedule**	**Total hours**	**Schedule**
Workday	10	9–16 18–21	9	5–9 16–18 21–24	5	0–5
Saturday	2	11–12 19–20	16	6–11 12–19 20–24	6	0–6
Sunday or holiday	0	–	8	11–13 17–23	16	0–11 13–17 23–24

Source: Resolución CREE 073, Comisión Reguladora de Energía Eléctrica. Obtenido de Resolución CREE 073, 20 de Abril de 2016: https://www.cree.gob.hn/wp-content/uploads/2019/02/REGLAMENTO-PARA-EL-CALCULO-DE-TARIFFS-PROVISIONALES.compressed.pdf.

The tariff is divided into four components: generation, transmission, distribution, and commercialization, each of which represents a cost to be covered, and its sum represents the tariff (Fig. 4.3).

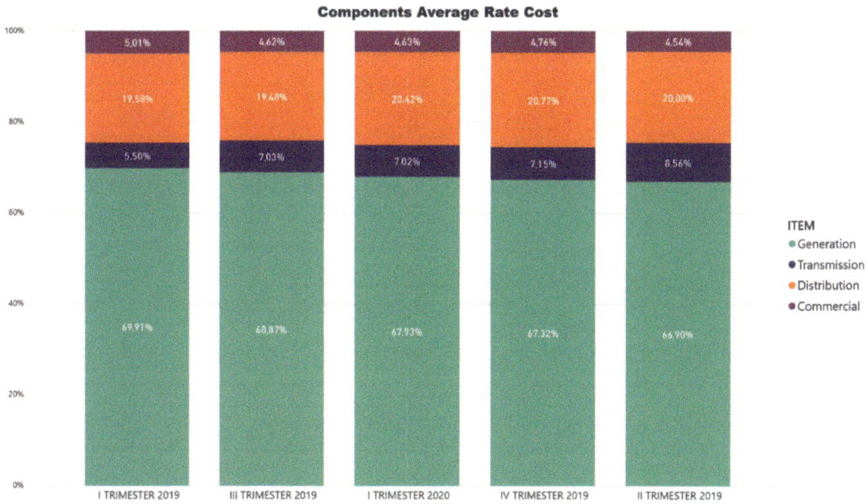

Figure 4.3 Average tariff cost components.
Source: Historial de Tariffs CREE, Comisión Reguladora de Energía Eléctrica. Obtenido de HISTORIAL DE TARIFFS CREE, Diciembre de 2019: https://www.cree.gob.hn/historial-de-Tariffs/; Informe de Ajuste Tariff Primer Trimestre 2020, Comisión Reguladora de Energía Eléctrica. Obtenido de Informe de Ajuste Tariff Primer Trimestre 2020, Enero de 2020: https://www.cree.gob.hn/informe-trimestral-de-Tariffs/.

The current fixed tariff depends on two important components, the prices of international fuels and the price of the USD against the Lempira. To estimate the price of the current hourly tariff, an estimated analysis was made regarding the price of the dollar over the last few years.

5.3 Correlation between the electric tariff and the dollar exchange tariff

The analysis considers the hourly tariffs based on quarters of the monthly average sale, from January 2016 to December 2019. The prices established for the hourly tariff in 2016 are considered as shown in Table 4.5.

The price of the dollar with respect to the Lempira is analyzed quarterly, as shown in Table 4.6.

The table uses the equivalent to the Honduran Lempira (L) in relation to the US Dollar ($).

From the above analysis, the hourly tariff proposal for quarter I of the year 2020 is obtained as shown in Tables 4.7–4.9.

The average sale of the dollar in the Honduran financial system was obtained calculating the dollar price monthly, as shown; this comparison was made between Tables 4.7 and 4.9 of the present chapter for the peak, middle, and valley periods.

Table 4.5 Tariff schedule.

Service	Fix rate	Power cost	Energy cost		
			Peak	Middle	Valley
	L/kWh	L/kWh	L/ kWh	L/ kWh	L/ kWh
Consumption >500 kWh/ month	54.57	177.2066	4.2017	2.184	1.8444
Low-power general service	54.57	200.0657	3.9883	2.0731	1.7507
Middle-power service	2280	246.55	3.5377	1.8575	1.5805
High-power service	5700	206.29	3.367	1.7131	1.5203

Source: Resolución CREE 073, Comisión Reguladora de Energía Eléctrica. Obtenido de Resolución CREE 073, 20 de Abril de 2016: https://www.cree.gob.hn/wp-content/uploads/2019/02/REGLAMENTO-PARA-EL-CALCULO-DE-TARIFFS-PROVISIONALES.compressed.pdf.

Table 4.6 Average dollar selling price 2016–20.

Average sale price of the dollar in the financial system Monthly series 2016–20 (Lempiras per USD $1)					
Month	**2016**	**2017**	**2018**	**2019**	**2020**
January	22.6120	23.7525	23.7641	24.5076	24.8221
February	22.7456	23.7323	23.7288	24.5542	24.8441
March	22.7845	23.6964	23.7816	24.5939	24.8882
April	22.7484	23.6266	23.8170	24.6041	24.9795
May	22.7662	23.6345	23.9682	24.6287	
June	22.9009	23.6078	24.1202	24.6769	
July	22.9993	23.5990	24.1541	24.6710	
August	23.0492	23.5360	24.1656	24.7124	
September	23.1264	23.5464	24.2076	24.7778	
October	23.2431	23.6304	24.2426	24.8074	
November	23.3427	23.7332	24.3920	24.8093	
December	23.6202	23.7229	24.4989	24.8179	
Average	**22.9949**	**23.6515**	**24.0701**	**24.6801**	**24.8835**

Source: Banco Central de Honduras, Tipo de Cambio. Obtenido de Banco Central de Honduras, 14 de Mayo de 2020: https://www.bch.hn/tipo_de_cambiom.php.

The calculation to determine each of the periods that comprise the hourly tariff is the following: Energy Price Medium Voltage Service 2016 times the average which equals the Energy Price in Dollars for May 2016 to December 2019; this result is

Table 4.7 Tariff proposed schedule.

	Fix rate	Power cost	Energy cost		
			Peak	Middle	Valley
Service	**L/kWh**	**L/kWh**	**L/kWh**	**L/kWh**	**L/kWh**
Consumption >500 kWh/month	L 54.57	L 212.65	L 4.5874	L 2.3845	L 2.0137
Low-power general service	L 54.57	L 240.08	L 4.3544	L 2.2634	L 1.9114
Middle-power service	L 2280.00	L 295.86	L 3.8624	L 2.0280	L 1.7256
High-power service	L 5700.00	L 247.55	L 3.6761	L 1.8703	L 1.6599

Source: Own elaboration.

multiplied with the dollar exchange rate per quarter, in order to obtain the Peak Period Medium-Voltage Service.

A relation is found between the energy in the first quarter of 2016, and the dollar price in that same quarter, then compared with the growth or decrease as shown Table 4.8.

Fig. 4.4 shows the projection of the approximate cost of the hourly tariff in the peak period from 2016 to 2020 (Table 4.8).

Fig. 4.5 shows the approximate projection of the cost of the hourly tariff in the intermediate period through time from 2016 to 2020 (Table 4.9).

In 2016, the hourly tariff cost was published. For the calculation of the cost of electric energy in the hourly tariff for the peak, middle, and valley periods of 2019, the sale price of the dollar was considered quarterly from January 2016 to December 2019, to give a linear comparison of growth and thus have an approximate projection of the new cost of the hourly tariff for the year 2019 (Fig. 4.6 and Table 4.10).

5.4 Load profiles and consumption sectors in Honduras

The energy consumption subsectors in Honduras are divided as shown in Table 4.11.

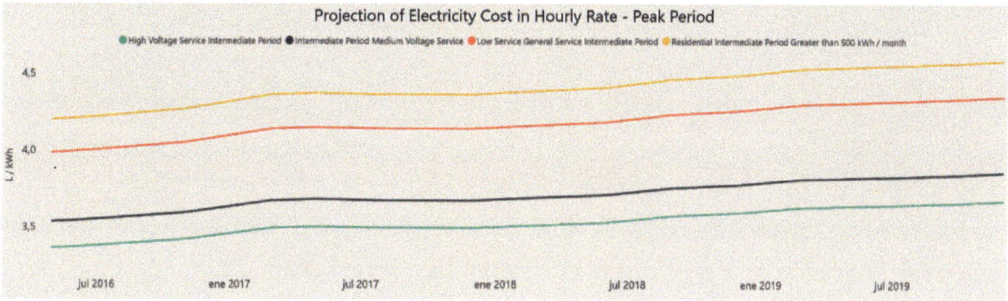

Figure 4.4 Projection of energy cost in the peak period 2016–20.

Table 4.8 Calculation of the energy price for the medium-voltage peak period 2016–20.

Evaluated quarter month	Average sale price of the dollar per quarter	Energy price medium-voltage service (2016)	Energy price in dollars (May 2016)	Peak period medium-voltage service	Price difference (2020 – 16)
May 2016	22.72262395	3.5377	0.1557	3.5377	0.0000
July 2016	22.83356462		0.1557	3.5550	0.0173
November 2016	23.10449303		0.1557	3.5972	0.0595
March 2017	23.61193373		0.1557	3.6762	0.1385
May 2017	23.66147565		0.1557	3.6839	0.1462
August 2017	23.61377583		0.1557	3.6764	0.1387
December 2017	23.61151464		0.1557	3.6761	0.1384
March 2018	23.73855939		0.1557	3.6959	0.1582
June 2018	23.85561197		0.1557	3.7141	0.1764
September 2018	24.14663363		0.1557	3.7594	0.2217
December 2018	24.28072517		0.1557	3.7803	0.2426
March 2019	24.52026033		0.1557	3.8176	0.2799
July 2019	24.62588711		0.1557	3.8340	0.2963
October 2019	24.72040571		0.1557	3.8487	0.3110
December 2019	24.80836537		0.1557	3.8624	0.3247

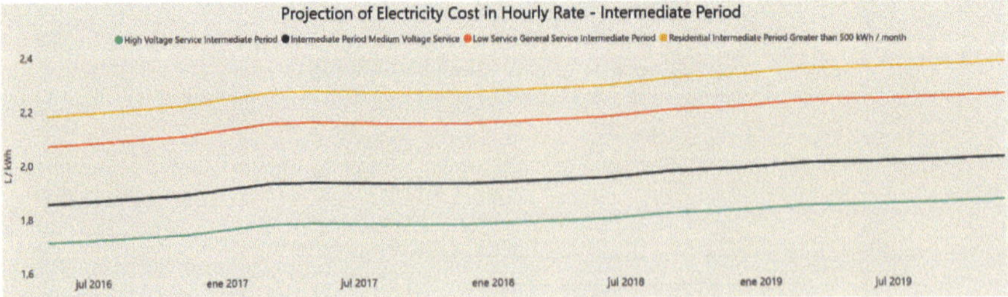

Figure 4.5 Projection of energy cost in the intermediate period 2016–20.

Table 4.9 Calculation of the energy price for the medium-voltage intermediate period 2016–20.

Evaluated quarter month	Average sale price of the dollar per quarter	Energy price medium-voltage service (2016)	Energy price in dollars (May 2016)	Peak period medium-voltage service	Price difference (2020 – 2016)
May 2016	22.72262395	1.8575	0.0817	1.8575	0.0000
July 2016	22.83356462		0.0817	1.8666	0.0091
November 2016	23.10449303		0.0817	1.8887	0.0312
March 2017	23.61193373		0.0817	1.9302	0.0727
May 2017	23.66147565		0.0817	1.9342	0.0767
August 2017	23.61377583		0.0817	1.9303	0.0728
December 2017	23.61151464		0.0817	1.9302	0.0727
March 2018	23.73855939		0.0817	1.9405	0.0830
June 2018	23.85561197		0.0817	1.9501	0.0926
September 2018	24.14663363		0.0817	1.9739	0.1164
December 2018	24.28072517		0.0817	1.9849	0.1274
March 2019	24.52026033		0.0817	2.0045	0.1470
July 2019	24.62588711		0.0817	2.0131	0.1556
October 2019	24.72040571		0.0817	2.0208	0.1633
December 2019	24.80836537		0.0817	2.0280	0.1705

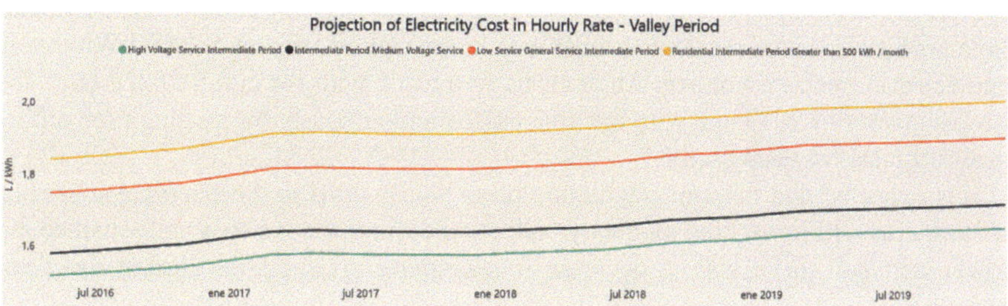

Figure 4.6 Projection of energy cost in the valley period 2016–20.

Table 4.10 Calculation of the energy price for the medium-voltage valley period 2016–20.

Evaluated quarter month	Average sale price of the dollar per quarter	Energy price medium-voltage service (2016)	Energy price in dollars (May 2016)	Peak period medium-voltage service	Price difference (2020 – 2016)
May 2016	22.72262395	1.5805	0.0696	1.5805	0.0000
July 2016	22.83356462		0.0696	1.5882	0.0077
November 2016	23.10449303		0.0696	1.6071	0.0266
March 2017	23.61193373		0.0696	1.6424	0.0619
May 2017	23.66147565		0.0696	1.6458	0.0653
August 2017	23.61377583		0.0696	1.6425	0.0620
December 2017	23.61151464		0.0696	1.6423	0.0618
March 2018	23.73855939		0.0696	1.6512	0.0707
June 2018	23.85561197		0.0696	1.6593	0.0788
September 2018	24.14663363		0.0696	1.6795	0.0990
December 2018	24.28072517		0.0696	1.6889	0.1084
March 2019	24.52026033		0.0696	1.7055	0.1250
July 2019	24.62588711		0.0696	1.7129	0.1324
October 2019	24.72040571		0.0696	1.7195	0.1390
December 2019	24.80836537		0.0696	1.7256	0.1451

Table 4.11 describes the number of users with the option of being billed, as per their level of consumption, through an hourly tariff. In this sense, it is proposed to take two load profiles from the industrial tariff, two load profiles from the commercial tariff, and three profiles from the residential tariff with consumption stratum >2000 kWh that is connected in medium voltage. An analysis was made with the current fixed tariff for the last quarter of 2019 and with the time tariff simultaneously by creating time difference formulas for each sector.

It is expected that the implementation of an hourly tariff in the electrical subsector of Honduras will impact the monetary collection during peak periods and sensitize the power demand curve with a decrease in electricity consumption during the peak periods.

Using the cost per energy shown in Table 4.5 for the fixed rate, the proposed cost per energy shown in Table 4.6 for the hourly rate, and the parameters of hourly blocks

Table 4.11 Number of qualified users by consumption stratum for hourly tariff.

Tariff/stratum classification	Total users per tariff/stratum
Residential	**63,988**
501–750	36,927
751–1000	12,536
1001–2000	11,591
>2000	2934
Commercial	**37,859**
501–750	10,122
751–1000	5858
1001–2000	10,449
>2000	11,430
Industrial	**1228**
501–750	79
751–1000	54
1001–2000	140
>2000	955
Total general	**103,075**

Source: Self-made.

defined in Table 4.4, the following correlative graphs between the two types of rates were elaborated, employing the Microsoft BI software to obtain the visualization of energy costs for each period of time, evaluating the residential, commercial, and industrial subsectors with a consumption stratum greater than or equal to 2000 kWh.

The software used allows the generation of a formula with which the distinction is made by time period, assigning a cost to the electrical energy consumed for each hour of the day and identifying the day (business days, Saturday and Sunday, or holiday).

The following figures show the comparative cost of energy charges with the fixed rate in effect for the first quarter of 2020 and the proposed hourly rate for the industrial, commercial, and residential sectors.

The graphs above represent the comparative charge in each of the sectors (industrial between Figs. 4.7 and 4.10, commercial between Figs. 4.11 and 4.14, and the

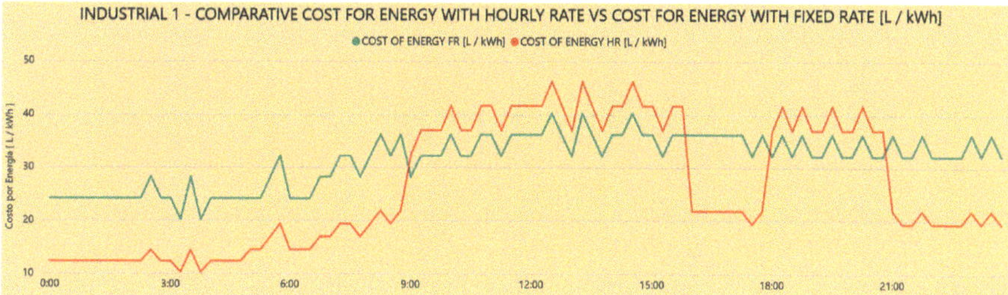

Figure 4.7 Comparative cost for energy of the industrial sector with an hourly tariff versus the cost for energy with a fixed tariff.

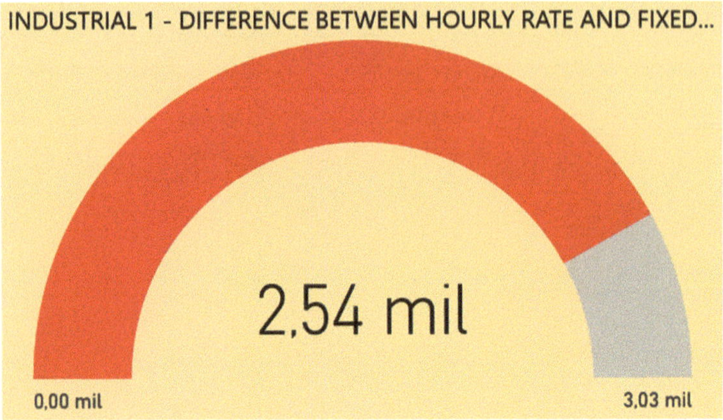

Figure 4.8 Collection of energy represented in Lempiras of the industrial sector.

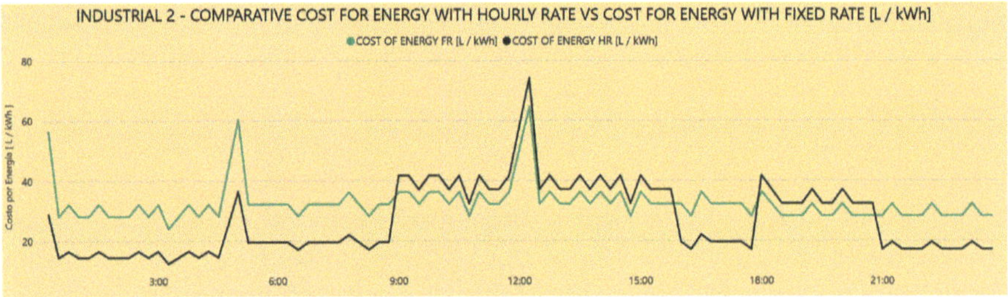

Figure 4.9 Comparative cost for energy of the industrial sector, with an hourly tariff versus the cost for energy with a fixed tariff.

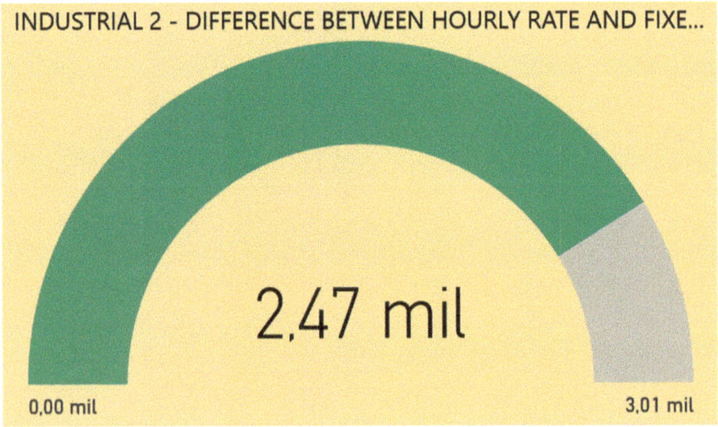

Figure 4.10 Energy collection represented in Lempiras of the industrial sector.

residential sector between Figs. 4.15 and 4.20) during one day of the first quarter of 2020 with the fixed tariff and the proposed hourly tariff. The purpose is to show the behavior of the load profiles when the billing per hourly blocks is implemented.

In the industrial sector, it is observed that the higher consumption peaks, during the day shifts in peak and middle hourly periods, represent higher costs with the proposed hourly tariff. The latter is the incentive that the distributor company can use to recommend a behavioral change in the electric energy consumption. In Figs. 4.7 and 4.9 it is observed that the valley periods with a reduction in the cost per energy consumption, being attractive to the industries, the production transfer a that hours, clearly more studies most exist in respect of the factors that intervene in the industrial production. Figs. 4.8 and 4.10 show the shaded area as the total value to be paid if the hourly tariff and being the limit of this graph the paid value per concept of energy for the in-force tariff in the first quarter of 2020.

The commercial sector is a wide market of clients that comprise different types of demand curves. In Figs. 4.11 and 4.13 are shown the variety of peaks that this demands without having a similarity between both analyses. Through the software Power BI the information of the load profiles was gathered and the cost per concept of energy were compared, as well as the fixed tariff and hourly tariff, for benefit the users, who will

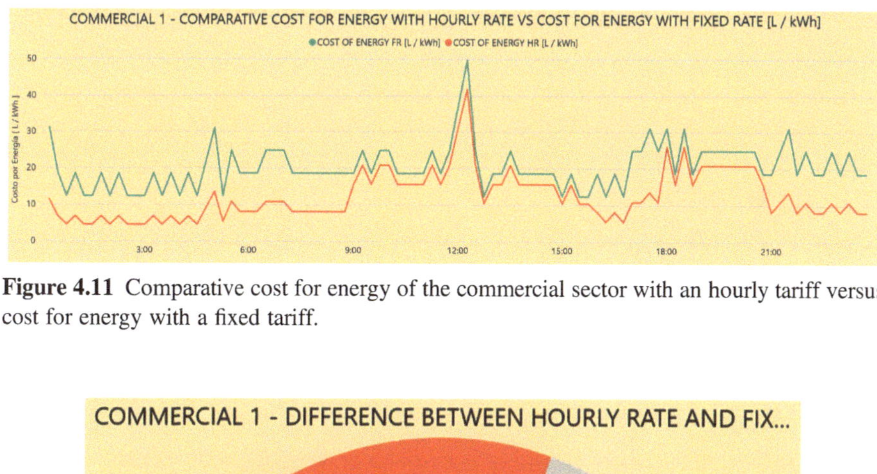

Figure 4.11 Comparative cost for energy of the commercial sector with an hourly tariff versus cost for energy with a fixed tariff.

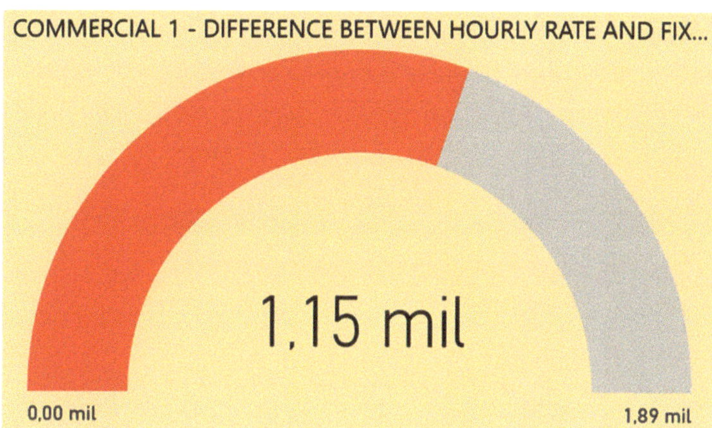

Figure 4.12 Collection of energy represented in Lempiras of the commercial sector.

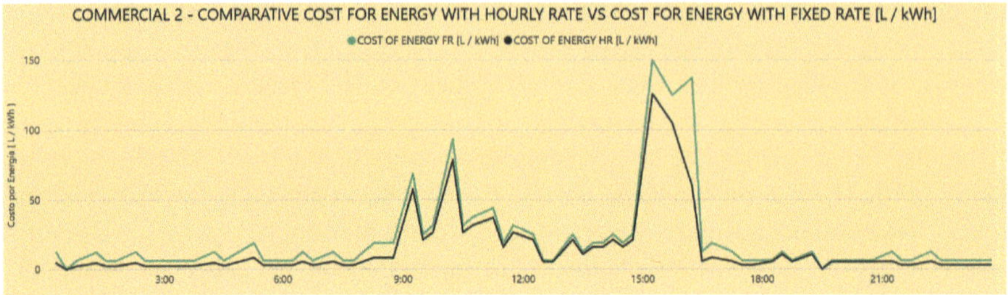

Figure 4.13 Comparative cost per energy of the commercial sector with an hourly tariff versus the cost for energy with a fixed tariff.

pay an equal or lower amount due the same demand behavior. This will harm the distribution company due the weaknesses of collection in Lempiras and the energy demand.

The residential subsector is even more variable than the commercial subsector; in this document, three figures for behavioral load profiles of different electrical demands are shown. In Fig. 4.15 is shown the nightly demand, resulting in a benefit for the final

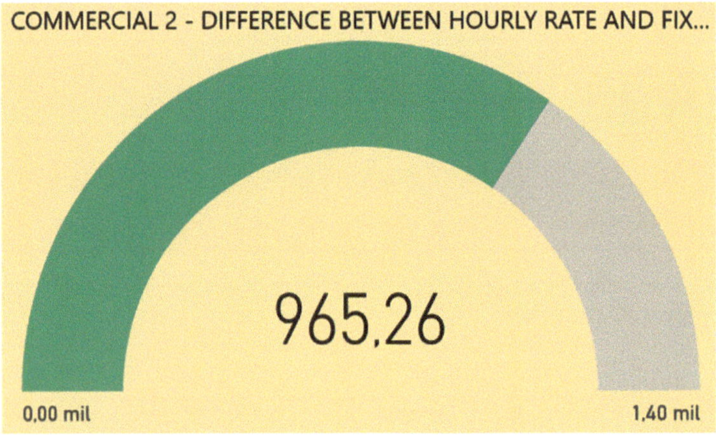

Figure 4.14 Energy collection represented in Lempiras of the commercial sector.

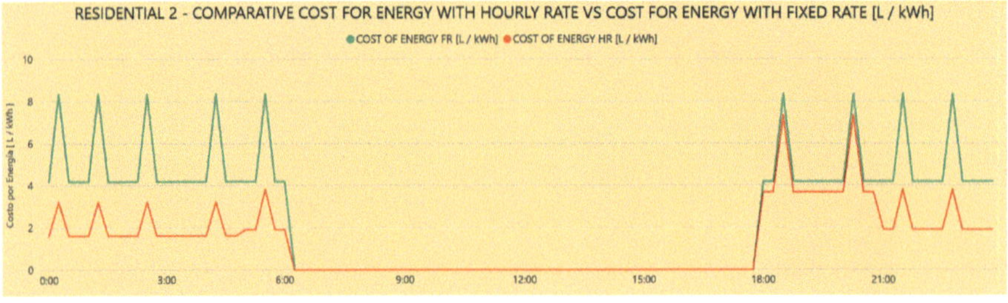

Figure 4.15 Comparative cost per energy use of the residential sector with an hourly tariff versus cost for energy with a fixed tariff.

user, with the implementation of an hourly tariff, because this demand is concentrated greatly in the valley periods (where cost per electric energy is cheaper); Fig. 4.17 is a demand profile with different peaks, during the evaluated 24 h of the day, and even so, the final user benefits with the hourly tariff, reducing his costs without changing his behavior. And a similar situation is seen in Fig. 4.19, in which the final user does not have to change his consumption behavior of electrical energy (Fig. 4.20).

The latter was analyzed through two load profiles for the industrial and commercial subsectors, and three load profiles for the residential subsector. Each one is compared with the first quarter of 2020 in-force fixed tariff and the proposed hourly tariff for the same amount of time.

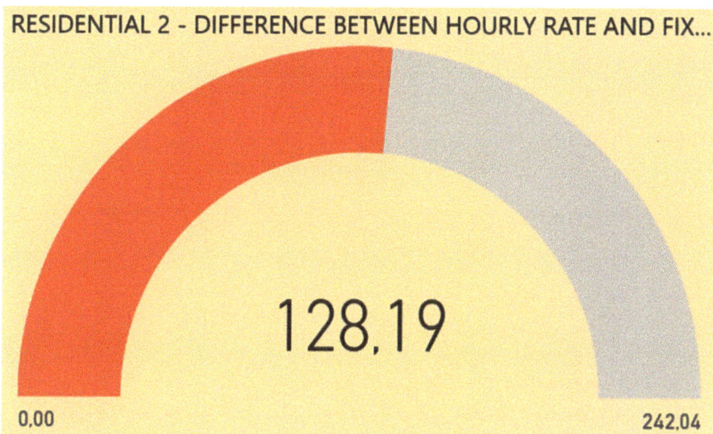

Figure 4.16 Collection of energy represented in Lempiras of the residential sector.

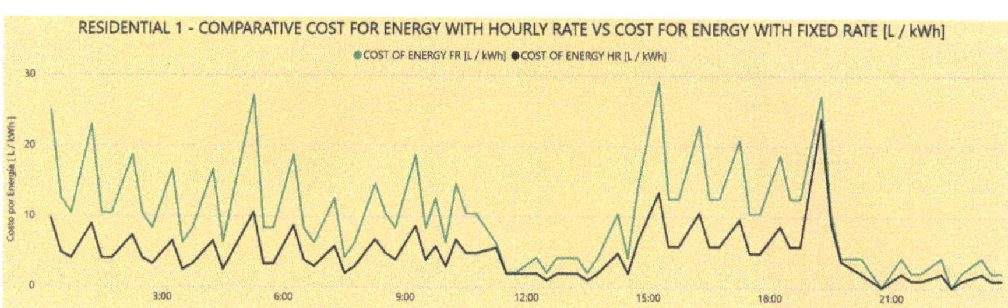

Figure 4.17 Comparative cost for energy of the residential sector with an hourly tariff versus the cost for energy with a fixed tariff.

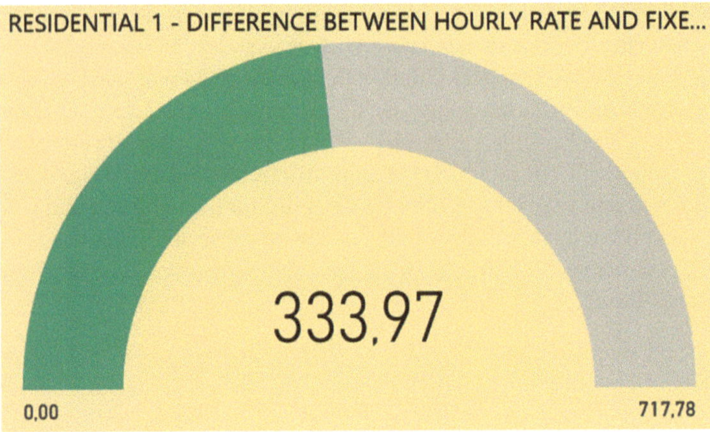

Figure 4.18 Comparative cost for energy of the residential sector with an hourly tariff versus the cost for energy with a fixed tariff.

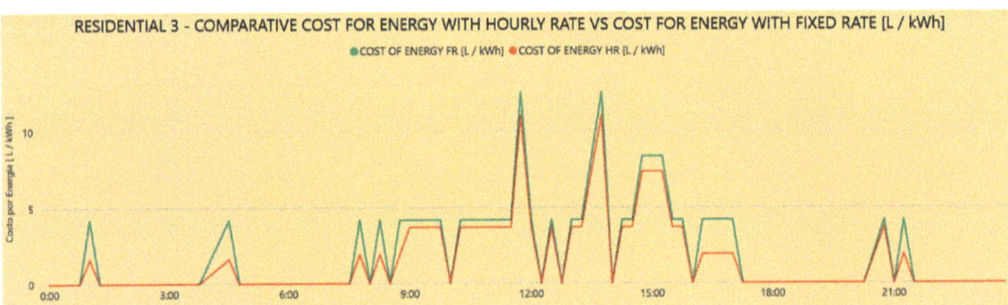

Figure 4.19 Comparative cost for energy of the residential sector with an hourly tariff versus the cost for energy with a fixed tariff.

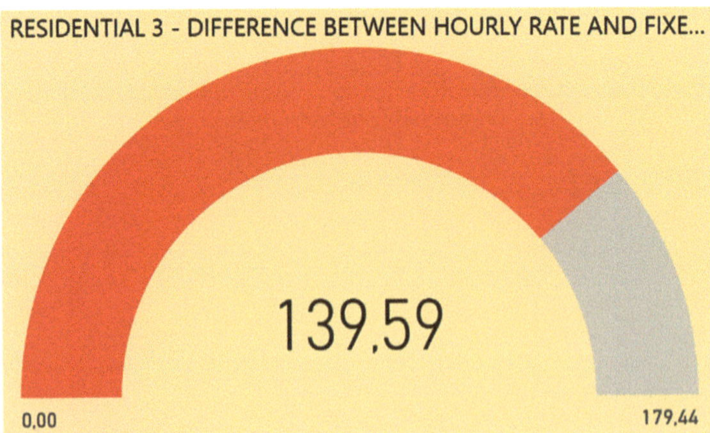

Figure 4.20 Comparative cost for energy of the residential sector with an hourly tariff versus the cost for energy with a fixed tariff.

6. Conclusions

1. Due to the large number of users that make up the electricity demand curve in the electricity subsector of Honduras, this study was carried out using load profiles from the tariff sectors: industrial, commercial, and residential, whose energy consumption stratum is over 2000 kWh to segment it as a viable scenario for the implementation of an hourly tariff.
2. By applying a correlative study between the hourly electricity tariff established in 2016 and the dollar exchange rate, the hourly electricity tariff for the first period of 2020 is obtained, corresponding to 3 months (January, February, March), and as a result, an increasing price variation was found of between 16% and 22% compared to the price published by CREE in 2016.
3. Different scenarios are established for the industrial, commercial, and residential subsectors of the Honduran electricity sector through its consumption stratum; with the case of the industrial sector >2000 kWh being the only one that appears to be viable in terms of a significant reduction in the cost of energy in hours of less activity. The cost for energy consumption during the peak period is higher than the current fixed tariff, causing a possible behavioral change during this period if the hourly rate is implemented in the monthly billing.
4. Through the analysis of the industrial, commercial, and residential subsectors greater than 2000 kWh and given the behavior shown in Figs. 4.9, 4.10, 4.11, 4.12, and 4.13, a reduction of the cost for energy consumption in the intermediate and valley hour periods can be obtained, and thus a benefit shown in the periods mentioned for energy consumption. Additionally, the electricity demand curve is reduced by a behavioral change in the consumption of users under study during peak periods.
5. Monthly savings are obtained in the electricity bill for the benefit of the end user as it is detailed the following table.

Sector	Figure No.	Evaluated date	Day savings evaluation (L)	Monthly savings (L)	Monthly savings ($)
Industrial 1	5.6	January 6, 2020	L 489.43	L 14,682.90	$ 587.12
Industrial 2	5.8	January 16, 2020	L 579.54	L 17,386.20	$ 695.22
Commercial 1	5.10	January 16, 2020	L 747.44	L 22,423.20	$ 896.63
Commercial 2	5.12	February 27, 2020	L 437.26	L 13,117.80	$ 524.54
Residential 1	5.14	February 6, 2020	L 113.85	L 3415.50	$ 136.57
Residential 2	5.16	February 15, 2020	L 383.81	L 11,514.30	$ 460.42
Residential 3	5.18	February 20, 2020	L 72.07	L 2162.10	$ 86.46

For an average of 30 days of monthly billing and the dollar exchange rate as of May 25, 2020, with a correlation between the tariff (fixed and hourly) under study and the load profiles of each subsector (residential, commercial, and industrial), monetary savings are obtained in each analysis performed; however, this does not apply to the distribution company.

7. Recommendations

The following are the recommendations arising from this study:

1. Performing the evaluation of energy consumption by sectors to determine a smaller population for the implementation of hourly tariff considering, in addition to the consumption stratum >500 kWh, the type of measurement technology, signal coverage area of smart meters (telemetry), and the feasibility of hourly tariff implementation in the main cities of Honduras (Tegucigalpa, Comayagüela, San Pedro Sula).
2. In conjunction with the liberalization of the electricity market in the country, there needs to be a study that establishes the type of generation technology available at peak, intermediate, and semivalid hours that will be responsible for supplying demand during those hours, in order to establish a tariff price as per the type of power generation technology (both established and supported).
3. Before the implementation of the hourly tariff, it is recommended to undertake training to understand the schedules established as peak, intermediate, and valley periods, to achieve a change in the energy demand behavior of users in the sectors that apply to this type of charging.

References

[1] AETS en asosiación con BCEOM y EDE Ingenieros, Tarifas multihorarias, socializacion e impacto - GAUREE II, Tegucigalpa M.D.C.: EuropeAID, 2011.

[2] Banco Central de Honduras, Tipo de Cambio, Obtenido de Banco Central de Honduras:, 14 de Mayo de 2020. https://www.bch.hn/tipo_de_cambiom.php.

[3] Comunicado CREE 01 2016, Comisión Reguladora de Energía Eléctrica, Obtenido de Comunicado CREE 01 2016, 31 de Mayo de 2016.

[4] ENEE Bolentin Estadístico Abril, ENEE, Obtenido de ENEE, Abril de 2019. http://www.enee.hn/planificacion/2019/Junio/Boletin%20Estadistico%20Abril%202019.pdf.

[5] Historial de Tariffs CREE., Comisión Reguladora de Energía Eléctrica, Obtenido de HISTORIAL DE TARIFFS CREE:, Diciembre de 2019. https://www.cree.gob.hn/historial-de-Tariffs/.

[6] Informe de Ajuste Tariff Primer Trimestre 2020, Comisión Reguladora de Energía Eléctrica, Obtenido de Informe de Ajuste Tariff Primer Trimestre 2020 :, Enero de 2020. https://www.cree.gob.hn/informe-trimestral-de-Tariffs/.

[7] Observatorio de la Energía UNITEC HN, Observatorio de la Energía UNITEC HN, Obtenido de Gracia Pineda:, 04 de Agosto de 2019. https://public.tableau.com/profile/gp17447#!/vizhome/EstadsticasElectricidadHonduras/EstadsticasENEE.

[8] Resolución CREE 073, Comisión Reguladora de Energía Eléctrica, Obtenido de Resolución CREE 073:, 20 de Abril de 2016. https://www.cree.gob.hn/wp-content/uploads/2019/02/REGLAMENTO-PARA-EL-CALCULO-DE-TARIFFS-PROVISIONALES.compressed.pdf.
[9] Resolución CREE 148, Comisión Reguladora de Energía Eléctrica, Obtenido de Comisión Reguladora de Energía Eléctrica:, 24 de Junio de 2019. https://cree.gob.hn/wp-content/uploads/2019/02/Reglamento-de-Tariffs.pdf.

Solar energy and gasification of MSW: two promising green energy options

5

Soumitra Mukhopadhyay
Doosan Power Systems India Pvt. Ltd., Kolkata, West Bengal, India

1. Introduction

Energy generated for electricity, process heating, and other applications is classified into two main categories: (i) renewable energy (RE) and (ii) nonrenewable energy (NRE). Irrespective of the country, worldwide, nonrenewable energy sources share a major stake in the electricity generation, but producing energy from these nonrenewable sources has a severe impact on our environment, polluting air, land, and water. On the other hand, electricity can be produced from various RE options with fewer impacts on the environment. A major advantage of this option is that it is possible to generate electricity without emitting CO_2 (carbon dioxide), which plays a major role in global climate change. Nowadays, global climate change is a very serious issue on which scientists, researchers, academics, industrialists, and governments are pondering over. Climate change has adverse effects on ecosystems, oceans, humans, and the weather. The Paris Agreement, signed in 2016 within the United Nations Framework Convention on Climate Change (UNFCCC), had hoped that average temperature rise globally should be sufficiently below 2°C above the temperature at the start of the Industrial Revolution, as per Article 2 [1]. However, the electricity demand is increasing and, as per the prediction of the World Energy Council [2], electricity demand will reach its peak globally in 2030. To meet this energy demand, many countries still depend on conventional fossil fuels, which are not only polluting in nature but also depleting daily. This has resulted in an urgent need to find alternate sources of energy, i.e., renewable energy options.

It is a well-accepted fact that electricity demand can suitably be met by renewable energy options through sustainable energy supply and emissions are also reduced, while RE options are used. In recent years, many countries, developing as well as developed, have adopted sustainable policies to shift to renewable energy options, reducing the use of conventional energy sources, which it will lead to: (i) achievement of sustainable energy growth and (ii) mitigation of global climate change [3]. Now, almost all countries are promoting awareness among their citizens of saving energy through the utilization of renewable energy options like solar, biomass, wind, municipal solid waste (MSW), hydropower, etc. It is also evident that, nowadays, renewable energy options have become cheaper than before. Technology advancements to increase efficiency through research and development (R&D) along with proper

Green Energy Systems. https://doi.org/10.1016/B978-0-323-95108-1.00003-3

regulatory policies [4] and tax exemptions in the renewable sector will attract entrepreneurs for investment in the renewable energy sector, with every country attempting to conserve energy and the environment. This will also increase job opportunities at every level, unskilled, skilled, semiskilled, technicians, engineers, etc., and will benefit people, business, and mother Earth.

In this chapter, nonrenewable energy sources and their associated demerits are discussed, first including the issues of climate change, depletion of fossil fuels, etc. Then, various advantages of renewable energy options are discussed. Detailed discussion is provided on two promising options for renewable energy—solar energy and energy from MSW. Various models have been shown on how electricity can be generated from these two RE options.

2. Nonrenewable energy

A nonrenewable energy resource is a natural substance which cannot be replenished once it is consumed; it is finite in nature. The reserves of nonrenewable resources are diminished inexorably once they are extracted from nature and used for human benefit as their stocks or reserves cannot be regenerated or restored.

Fossil fuels like natural gas, coal, and oil are mainly considered as nonrenewable energy sources. Other than these fuels, nuclear fuels also belong to this category. Animals and plants, buried under the ground several millions of years ago, i.e., in the Carboniferous period, were transformed into fossil-based fuels. These animals and plants can be classified into two groups: plankton and plants. While plants have been converted into coal, plankton has been transformed into oil and natural gas. Today, coal is extracted through coal mining and oil and gas are extracted through the drilling of wells on land and offshore. Therefore, it is a very slow geological transformation process of organic matter into hydrocarbons and other organic compounds under high pressure and temperature, and low oxygen. As fossil fuels are still being produced today, it may be argued that these can be treated as a renewable source. However, one has to keep in mind about the extraction to regeneration rate; these fossil fuels are extracted and used at a faster rate than the rate of regeneration, which is very slow. Therefore, fossil fuels can only be treated as nonrenewable.

Till now, a major portion of total primary energy consumption is met by nonrenewable energy options all over the world, as shown in Table 5.1 [5]. It is evident that more than 80% of the energy demand is met by natural gas, coal, and oil, but these sources have a lot of hazardous impacts on the environment.

2.1 Demerits of nonrenewable energy

Although fossil fuels meet a major portion of energy demand all over the world, they have various hazardous impacts, which are described below in brief:

(A) Environmental hazards: One of the major disadvantages of fossil fuels is the environmental pollution. To extract stored energy from fossil fuels, they need to be burned converting

Table 5.1 Total primary consumption by renewable and nonrenewable energy in 2019 [5].

Type of energy resource	Consumption % in 2019
Coal	27
Oil	33.1
Natural gas	24.2
Hydro	6.4
Other renewables	5
Nuclear	4.3

chemical energy into heat energy, and this process of burning leads to smoke and gases containing harmful particles and greenhouse gases (GHGs) being released into the atmosphere. The emitted gas leads to air pollution and acid rain, along with various hazardous impacts on human life. Particulate matters form soot and nitrogen oxides (NOxs), reacting with sunlight, and create ground-level ozone and smog. These result in various diseases like respiratory damage, cancer, asthma, bronchitis, etc.

(B) Global warming: This is a that, due to burning of fossil fuels, CO_2 gas is released into the atmosphere and this CO_2 traps heat from the sun leading to a continuous increase in average global temperature and global warming. As a result, polar ice caps melt, low-lying areas are flooded, and natural habitats are also destroyed.

(C) Land degradation and impact on aquatic life: Extraction, processing, and moving of underground coal, oil, and natural gas take an enormous toll on landscapes and ecosystems. In this process, huge areas of forests or land are impacted which cannot be returned back to their original state. Also, wildlife leave the forests and migrate to less suitable habitats.

During transportation of oil through tanks or ships, oil tankers occasionally leak or ships sink, causing oil to be released into the seawater. This oil contains some toxic substances which cause severe impact, on the aquatic lifeforms. During transportation of crude oil by sea, spillage of oil can also result in hazardous impacts on aquatic life by reducing the oxygen content of the water.

(D) A nonrenewable resource: Even if fossil fuels did not have the above dangerous impacts on the environment, they cannot be relied upon forever since fossil fuels are nonrenewable by nature, which means fossil fuels need millions of years to form under the ground and while industrialization and demand for electricity are increasing at a very faster rate others sources are required.

3. Building a clean energy future through renewable energy

For the reasons enumerated in the above sections, human cannot rely on fossil fuels alone as a future energy resource. The alternate option is renewable energy sourcesal, though RE sources (except hydro) have a much lower energy return on investment (EROI) than fossil fuel sources [6]. However, in scaling up energy efficiency and renewable energy, sufficient progress has been made throughout the world over the past decades. Falling prices and advancements in renewable technologies are pushing the world toward the use of cleaner renewable energy options, such as MSW, solar, wind, biomass, hydro, etc. A decade ago, renewable energy technologies were predominately seen from the environmental aspect only; those who were only interested in shifting from conventional to RE for environmental reasons were encouraged, but today RE technologies are used to provide electricity, in transportation, heating and cooling, etc. It has now spread worldwide, and sustainable growth is only possible through renewables. In addition to their environmental benefits, renewable sources have proved to be an economic driver, creating jobs and stimulating new technological developments. Power produced from RE sources can be integrated with the existing grid.

It is evident from Table 5.2 that the percentage share of electricity from renewable energy sources has increased by 4.6% in 2009–2018 and 3.0% in 2000–2009. Renewable energy sources have the following inherent advantages:

(A) **Inexhaustible energy:** The first and foremost importance of renewable energy is in its name—it is renewable. While fossil fuels will be depleted after several years, the sun will still be shining; the wind will still be blowing. Even if potential constraints of renewable energy are accounted for, studies have repeatedly suggested that renewable energy can take care of a significant share of future electricity demand.
(B) **Less global warming:** Global warming is a very serious issue which mankind is facing today due to the burning of fossil fuels during electricity generation, and human activity is affecting our environment through the emissions of GHGs, particulate matters, etc. In contrast most renewable energy sources emit minimal to zero global warming emissions, keeping our environment cleaner.
(C) **Improved public health:** The harmful pollutants from natural gas and coal-fired power plants cause various health issues such as neurological problems, heart failures, breathing problems, carcinogenic and skin problems, and many other serious diseases. Renewable energy options, being a cleaner source of energy, do not create such serious health issues.
(D) **Supports job growth:** Job opportunities in the renewable energy field have increased considerably since the year 2012. It has the potential to quickly increase employment opportunities and economic development. While 11.5 million people were engaged globally in the field of RE in the year 2019, it is expected that renewable jobs will increase to nearly 30 million globally by 2030 [8], as per International Renewable Energy Agency (IRENA).
(E) **Energy independence:** Now, households and industries can generate their own electricity without depending on the coal-powered energy grid. Solar panels can be installed on the rooftop or on the sides of factories to make themselves self-reliant, which will reduce the electricity cost also on a long-term basis. Excess electricity produced can be stored in battery or sent back to the grid.

Table 5.2 Global renewable electricity generation as a percentage of total generation [7].

Year	Global renewable electricity as % of total generation	% Increase
2000	18.7	3.0% (2000–2009)
2009	21.7	
2018	26.3	4.6% (2009–2018)

4. Various types of renewable energy

There are various types of renewable energy options, some of which include:

(i) Solar
(ii) Wind
(iii) Tidal
(iv) Geothermal
(v) Ocean thermal energy conversion (OTEC)
(vi) Biomass
(vii) Municipal solid waste
(viii) Hydel, etc.

Of the various options available, as mentioned above, the following two important options are discussed in this chapter:

(i) Solar and
(ii) Municipal solid waste.

5. Solar energy as renewable energy

Solar power can directly be harnessed from the sun. It is increasingly popular for generating electricity worldwide since

(i) It is a very clean and green energy source.
(ii) It is available sufficient quantity almost all over the world where life exists.
(iii) The Earth receives approximately 1.8×10^{11} MW power from the sun [9], which is very large in quantity in comparison to the present consumption rate.

In addition to harnessing power from solar energy, it can also be used for heating and desalinating water. Present and future energy demand globally can be met by solar energy continuously. Until the 1970s, entrepreneurs were not interested in investments for the commercialization of solar power because it was expensive due to the primitive technologies available at that time. From the 1980s onwards, there has been a remarkable development in the technology for obtaining electricity from solar energy and there has been a focus on solar energy conversion around the world. There are mainly two technologies available for this conversion: (i) solar photovoltaic and (ii) concentrated solar power (CSP).

5.1 Solar photovoltaic

5.1.1 Working of a solar photovoltaic system

The photovoltaic effect is a process of converting light, i.e., photons, into electricity. Solar cells or photovoltaic (PV) cells are electronic devices where sunlight is directly

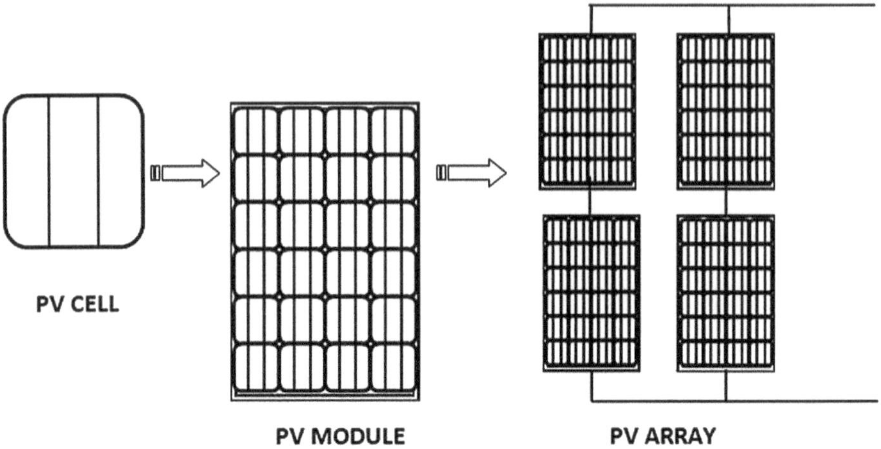

Figure 5.1 Photovoltaic hierarchy.

converted into electricity due to the photovoltaic effect. A photovoltaic system is an array of solar modules that comprise a number of solar cells that generate electrical power. Multiple modules, as shown in the photovoltaic hierarchy in Fig. 5.1, are wired together to form an array to obtain more electricity.

Each photovoltaic cell, which is a unit of a PV module, is basically a sandwich of two slices of semiconducting material, generally silicon. While the top layer is doped with pentavalent phosphorous, trivalent boron is doped in the bottom layer. The top layer produces extra electrons a having negative (−ve) charge and the bottom layer creates a positive (+ve) charge or holes. Therefore, an electric field is generated at the juncture of silicon layers. Whiwn incoming solar radiation, i.e., photons, strikes the photovoltaic cell, electrons are dislodged from the atoms. The electrons are pushed out of the silicon junction and travel to the front surface of the solar cell. Many electrons will move toward the front surface of the cell. This creates a charge imbalance or potential difference between the two surfaces, i.e., front and back, acting as two terminals of a battery. When an external load is applied between the two surfaces, electricity flows, as shown through an enlarged view in Fig. 5.2. Photovoltaic systems, being modular in nature, can be used to meet small or large power demands.

The electricity thus generated is DC, or direct current. An inverter is installed which is used to convert DC to AC, or alternating current. This alternating current (AC) is used to power all of the electrical appliances in households, factories, etc. This current can be stored in batteries and excess current can be transferred to the grid through metering. The PV modules rest on mounting structures which direct them toward the sun. A schematic arrangement of a photovoltaic power plant is shown in Fig. 5.2.

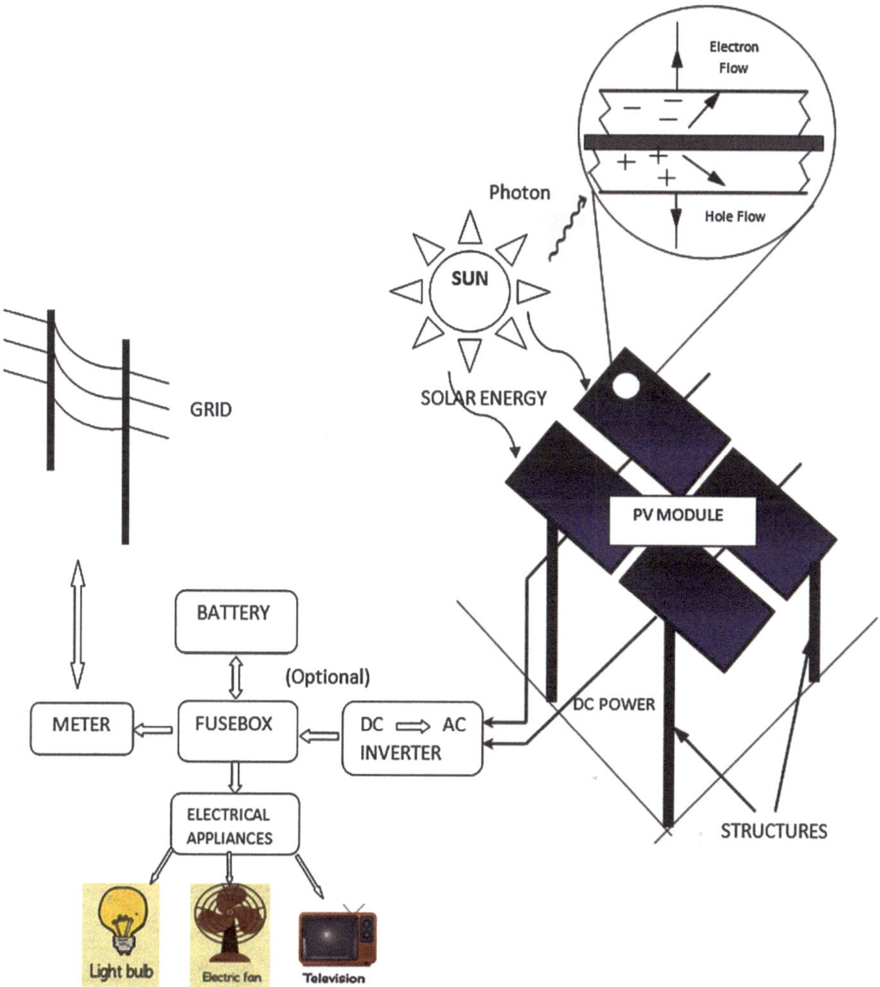

Figure 5.2 Schematic arrangement of a photovoltaic power plant.

5.1.2 Classifications of solar panels

Silicon-based PV panels can mainly be classified into three types:

(i) **Monocrystalline solar panels:** In monocrystalline solar panels, each cell contains a thin silicon wafer cut from a large single silicon crystal, and it is dark bluish black in color. It is the purest form of silicon. These panels are expensive and have an efficiency in the range of 17%–22%.

(ii) **Polycrystalline solar panels:** These are also thin wafers cut from a block of multiple silicon crystal. They are blue in color and have efficiencies in the range of 12%–15%.

(iii) **Amorphous silicon panels:** These panels are made of solar cells manufactured by depositing a thin layer of photovoltaic substances (such as amorphous silicon, cadmium telluride, copper indium gallium selenide, etc.) onto a solid surface, like glass. These cells are often called thin-film solar cells. These panels are less expensive and have an efficiency in the range 7%–10%.

5.1.3 Applications of a photovoltaic system

A solar photovoltaic system or PV system is an electricity generation system with a combination of various components such as PV panels, inverter, battery, mounting structures, etc. Nowadays, of the various renewable energy technologies available, PV is one of the fastest-growing renewable energy options. With the dramatic reduction of the manufacturing cost of solar panels, they will play a vital role in the generation of renewable power. Generally, these panels have a lifespan of roughly 25–30 years, and have various applications:

(i) **Utility applications:** The alternating current generated through an inverter, as shown in Fig. 5.2, can be used to power household devices and machines. The excess electricity can be sold to the grid also through a metering system. A PV system is coupled with a storage system, generally battery storage, from which power can be obtained during peak demand periods.

(ii) **Solar rooftop:** Today, adoption of solar rooftops is increasing due to their various advantages—cost effective, secure investment, support from government, reduction of carbon footprint, etc. For example, MNRE, i.e., Ministry of New and Renewable Energy, has targeted 40 GW solar rooftops in India by 2022 [10].

(iii) **Stand-alone systems:** Dependence on grid connections or utility is avoided in stand-alone systems and they can directly use the generated solar power. Battery storage can be used during low sunshine hours or night time. In some cases, conventional generators can be used as backup. The following are some applications of stand-alone systems:

   (A) **Solar lamps:** With the invention of the light-emitting diode (LED), PV systems find a suitable application in mobile or remote lighting systems. PV systems combined with battery storage are mostly used in public use facilities, billboards, lighting for trains, parking lots, etc.

   (B) **Electricity for remote areas:** Many areas are too distant from the distribution network to obtain power from the grid. Also, for many scientific experiments like seismic activities, meteorological information, etc., set-up is done on remote areas where a conventional grid power is not available. PV systems are a very attractive option for these cases.

   (C) **Disaster relief:** During natural calamities such as heavy storms, floods, earthquakes, etc., conventional power generation and distributed systems are often destroyed. In such situations, portable PV systems can provide temporary solutions for daily needs of electricity such as light, communication, food, and water systems. Emergency health clinics also prefer PV-based electricity over conventional systems.

   (D) **Solar water pumping:** A solar water pump system uses an electrical pump system in which the electricity is provided by photovoltaic (PV) panels. While sunlight falls on the solar panels, they produce direct current (DC) which is supplied to the motor of the pump. If the motor requires an alternating current (AC), the DC can be converted to AC electricity using an inverter. The water thus pumped out can be directly used in the field or can be stored in a suitable place for later use.

   (E) **Charging batteries of vehicles:** The batteries of electric vehicles can be charged at PV powered stations.

(iv) **Floating solar:** As the name suggests, a floating solar PV plant consists of a PV system mounted on structures floating on the surface of a body of water such as a lake, river, etc. The primary advantage of a floating system is that it avoids the need for valuable land area, which can otherwise be used for another purpose such as agriculture, etc.

5.1.4 Efficiency calculation of a solar cell

The efficiency of a solar cell (η) is defined mathematically by the following equation:

$$\eta = \frac{V_{OC} \times I_{SC} \times \text{Fill Factor}}{A_c \times G}$$
$$\text{Fill Factor} = \frac{\text{Maximum Actual Obtainable Power}}{V_{OC} \times I_{SC}} \tag{5.1}$$

where V_{OC} = voltage (open circuit), I_{SC} = current (short circuit), G = irradiance (W/m^2), and A_C = surface area in m^2.

The higher the fill factor, the better is the solar cell.

5.1.5 Disadvantages of a PV system

Though PV systems have various advantages, silicon from soil is depleted due to its use in PV cells and rare earth metals are also used in solar panels, which prohibit them from being the best solution to the power generation problem. A typical commercial solar cell has an efficiency of about 15%. Due to low efficiency, larger arrays are required, which leads to higher cost. The main demerits of solar PV cells are as follows:

(i) Limited efficiency levels
(ii) PV panels produce direct electric current (DC), and it requires an inverter to convert it into alternating current (AC), which is an expensive piece of electronic equipment.
(iii) PV cells cannot store excess energy. For a good PV system, it requires battery storage which can be used when sunlight is not available and this adds cost to the system.

5.2 Concentrated solar power

5.2.1 Working of a concentrated solar power

Most of the disadvantages of solar photovoltaics are overcome by CSP technology. In a CSP system, there are mirrors, sometimes in the hundreds of thousands, which track a large area of sunlight onto a receiver or absorber with a small aperture area. The concentrated solar heat is transferred to the working fluid or heat transfer fluid (HTF) in the receiver or absorber. The working fluid, having solar thermal energy, drives a heat engine or steam turbine, with which an alternator is coupled; thus, electricity is generated in a CSP system. The basic principle of electricity generation from a CSP system is shown through the simplified schematic in Fig. 5.3.

Archimedes, the Greek scientist, by the combined use of "burning glass" and bronze shields, concentrated solar energy that was focused on invading Roman wooden ships which were set on fire as a result. Thus, the Romans were defeated in theis battle in 212 BCE repelling them from Syracuse [11].

Figure 5.3 Basic principle of electricity generation from a CSP plant.

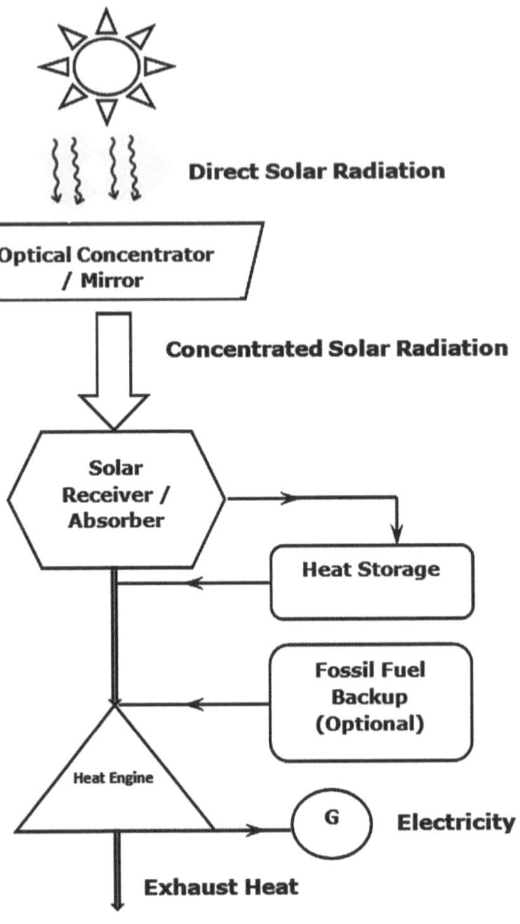

A single CSP plant can generate enough power to make it suitable for a utility-scale solar power market. The solar thermal energy thus generated is generally stored in a suitable medium which can be utilized to generate electricity during day or night or on cloudy days while sunshine is not available. While a CSP plant is backed up by large thermal storage facilities, it is expected that it can meet base, intermediate, and peak load demands by 2030, making it possible for low-carbon CSP plants to be available which could compete with conventional power plants that generate a very high level of GHG emissions. By 2050, CSPs are expected to provide 11.3% of global electricity with appropriate support such as fossil fuel or biomass, etc. [12,13]. Apart from traditional residential electricity generation, CSPs can be used in some other applications such as process heat, desalination, etc. Flat plate collectors, considered as non-concentrating collectors, have a concentration ratio equal to 1, which means that the heat collection area is equal to the area of concentration of radiation. Therefore, the low temperature of the working fluid is raised with this type of collector, which is not suitable for the power sector. However, if CSPs are used, working fluid can reach a high temperature due to higher concentration at the receiver/absorber. Using the Carnot cycle, the maximum theoretical thermal efficiency is given by the equation below:

$$\eta_{\text{thermal}} = \frac{T_H - T_L}{T_H} \tag{5.2}$$

where T_L is the minimum and T_H is the maximum temperature. From the above efficiency equation, it is evident that better efficiency of the heat engine is achieved, while working fluid with a higher operating temperature is used. Therefore, CSPs can be used for utility-scale power production since "higher quality" solar energy is collected in the receiver. As the solar energy is concentrated in a smaller area, heat losses due to convection, conduction, and radiation are also lower.

The main objectives of CSPs are as follows:

(i) To increase system efficiency through higher temperature operations
(ii) To minimize optical loss
(iii) To minimize thermal loss and,
(iv) To reduce the cost of the solar field.

5.2.2 Types of concentrated solar power

A concentrating solar power system can broadly be divided into single-axis and double-axis tracking systems, as shown in Fig. 5.4. Under a single-axis tracking system, the reflectors which are long and narrow in size rotate around a single axis and the solar radiation is focused on a linear receiver/absorber. In a dual-axis or double-axis tracking system, automated tracking technology is used for concentrators and it can track solar

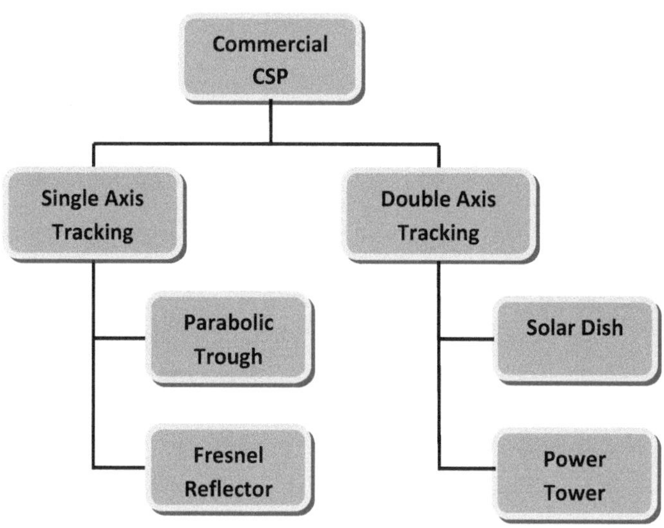

Figure 5.4 Broad classification of a commercial CSP plant.

radiation on a dual axis [14]. A dual-axis solar tracking system produces more power compared with single-axis tracking.

5.2.3 Parabolic trough system

The parabolic trough system is one of the most developed and commercially proven concentrated solar thermal power plants, and these have been operating for more than 25 years in various parts of the world. The parabolic trough power plant consists of the following major components:

(i) Metal support structure or concentrator structure
(ii) Parabolic reflectors or mirrors
(iii) Receiver tube and
(iv) Tracking system.

The above components are shown in Fig. 5.5. The concentrator support structure is the structural skeleton of the parabolic trough solar collector. It provides various functions in the system. The main functions of the structure are as follows:

(i) Supports the mirrors and receivers
(ii) Withstands wind and
(iii) Helps in tracking the sun.

The parabolic reflectors or concentrators are long, curved mirrors that focus sunrays into a tube, called an absorber. The parabolic reflectors track the sunrays around a linear axis, extending from north to south, during daytime, to follow the path of the sun from east to west. Concentrators can be continuous or segmented. The reflectors are generally composed of: (i) bent glass mirrors, (ii) silver-backed glass plates, or

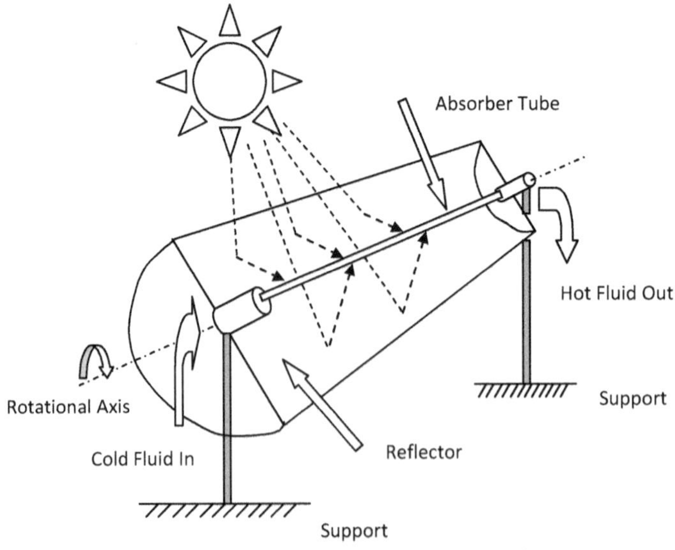

Figure 5.5 Major components of a parabolic trough power plant.

(iii) anodized aluminum sheets. They are widely available commercially. The absorber or receiver is kept along the focal line of the concentrator and an HTF, i.e., HTF like mineral oil, synthetic oil, or a molten salt, etc., runs through the tube. The absorber tube is made up of a several-meter-long metal tube and mostly a glass envelope covers it. For reduction of convective heat losses and allowing thermal expansion, the area between the tube and glass envelop is usually filled up with either air or a vacuum. The selective coating on the metallic tube has good absorptivity and a low thermal emittance for reducing thermal radiation losses.

The solar energy falls on the trough, is reflected back to the absorber tube, and heats up the HTF. The temperature of the HTF reaches about 400°C or more. Steam is produced in a heat exchanger, and the steam runs the steam turbine which produces power through a generator coupled with the steam turbine. Solar tracking helps in maximizing the use of concentrating collectors. If the solar tracking system is not used, the collector fields can be used for a very short period of sunshine hours of a day and the trough field cannot be used in optimum conditions. Generally, automated tracking technology with large concentrators is used nowadays to concentrate the sun on a biaxial path [14]. A solar thermal power plant with parabolic trough and a thermal storage system is shown schematically in Fig. 5.6. Some of the largest solar thermal power systems operating with parabolic trough technology are listed in Table 5.3.

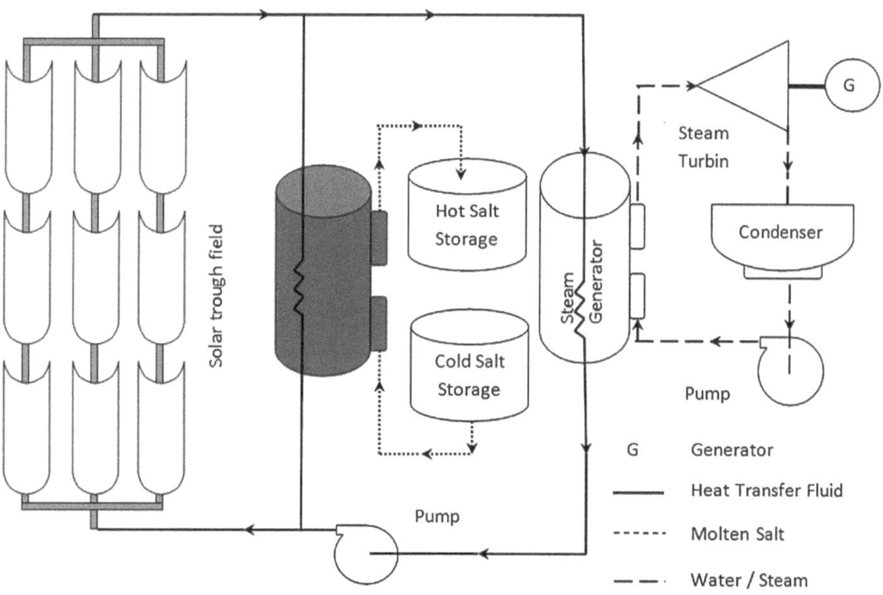

Figure 5.6 Solar thermal power plant with a parabolic trough and thermal storage system.

5.2.4 Linear Fresnel reflectors

Like parabolic troughs, solar radiation heats a receiver pipe in the case of linear Fresnel reflectors (LFRs), and the HTF fluid flows through the pipe. LFRs, which are based on

Table 5.3 Some of the largest parabolic trough power plants [15].

Name of power station	Location and country	Electrical capacity (MW)	Technology type
Noor Power Plant	Ghassate, Morocco	510	Parabolic trough along with central receiver system
Solar Energy Generating Systems (SEGS)	California, US	310	Trough technology
Mojave Solar Project	Barstow, California, US	280	Trough technology
Solana Generating Station	Gila Bend, Arizona, US	280	Trough technology
Genesis Solar Energy Project	Blythe, California, US	280	Trough technology

single-axis tracking technology, use long, thin segments of flat mirrors to focus sunlight onto a fixed absorber located at a common focal point of the reflectors and the absorber runs parallel to the reflector rotational axis. A schematic diagram of an LFR is shown in Fig. 5.7. These mirrors concentrate the sunlight to approximately 30 times its normal intensity to transfer the concentrated thermal energy to the thermal fluid located in the absorber to increase the temperature of the fluid. The thermal fluid is then transferred to the heat exchanger to heat the working fluid, which in turn runs a generator.

The working of an LFR is based on the effect of a Fresnel lens, which was first invented by a French physicist—Augustin-Jean Fresnel. It helps in the reduction of cost compared to parabolic trough reflectors since the lens are constructed with a large aperture area and short focal length. One major disadvantage with this construction is the shading effect of adjacent mirrors which can be overcome by utilizing more ground space but this increases costs. A 50-MW Fresnel reflector-based CSP plant with 15 h of thermal energy storage was completed in 2019 in Dunhuang, Gansu Province, in China [15].

5.2.5 Solar dish

Solar dish systems generally use parabolic or paraboloidal dish-shaped mirrors to focus the incoming solar radiation onto a receiver, positioned at the focal point of the dish. The dish concentrator is made of highly reflective, light-weight mirrors. The mirrors are kept on a rigid structure, which is mounted on a single truss support. The solar dish concentrates the incident beam irradiation onto a small region called the focal point, where the receiver is located, as shown in the schematic diagram in Fig. 5.8. The receiver absorbs the concentrated solar irradiation and the heat is transferred to the working fluid which flows inside the receiver. The solar dish continuously

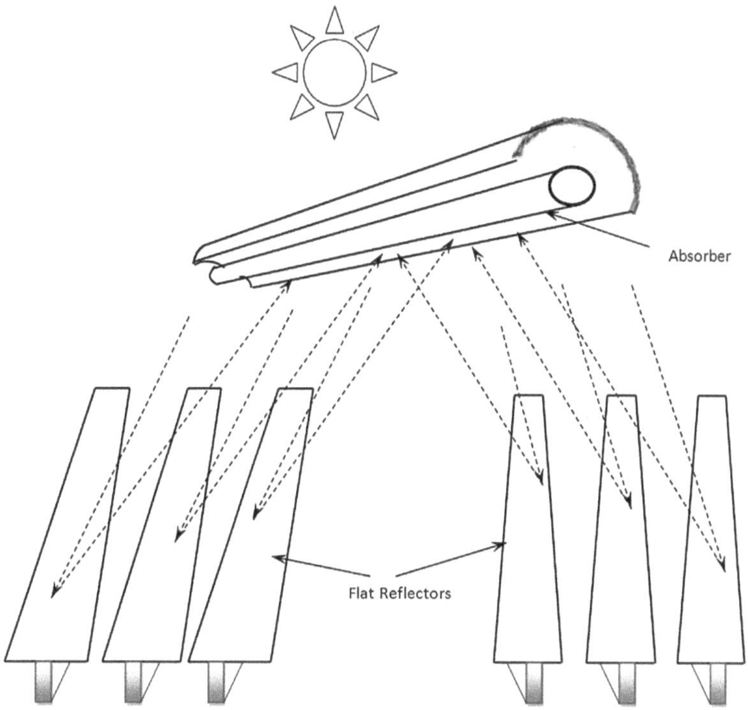

Figure 5.7 Schematic diagram of a linear Fresnel reflector (LFR).

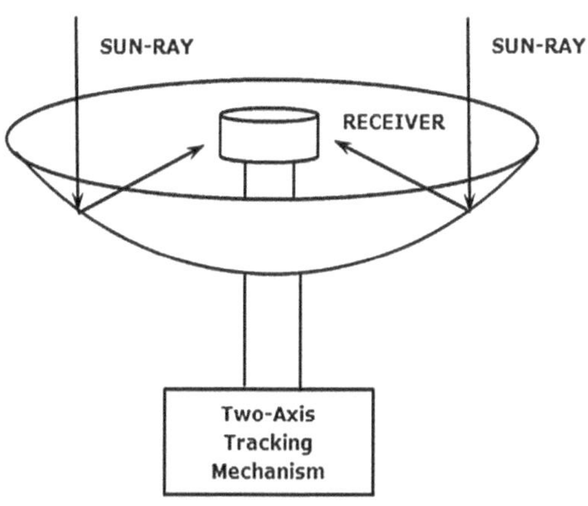

Figure 5.8 Schematic of a solar dish receiver system.

tracks the sun in two axes, while the parabolic trough tracks it in a single axis. The advantage of a dish system is that a higher concentration of light leads to much higher temperatures being achieved at the receiver working fluid. As already discussed, higher temperatures lead to better conversion to electricity, and from this point, the dish system is very efficient.

5.2.5.1 SAIC/STM solar dish system

Science Applications International Corporation (SAIC) jointly with Stirling Thermal Motors (STM) developed a dish-stirling system which consists of a faceted stretched-membrane dish, a direct illumination receiver, and a Stirling Thermal Motors' kinematic Stirling engine. The net output from the first-generation system is 21.6 kWe with 24% conversion efficiency at an insolation level of 1000 W/m^2 [16]. Another SAIC/STM joint venture project produced 30 kW power using an SAIC dish and an STM generation III Stirling engine operating at 720°C [17].

5.2.5.2 ARUN solar dish system

In India, Clique Developments Private Limited, Mumbai, in collaboration with IIT-Bombay, India, has developed ARUN70 and ARUN160 with an aperture area 70 and 160 m^2, respectively, which are Fresnel paraboloid concentrating solar collector system, with a two-axes tracking flat reflector dish having an insulated receiver with minimum thermal losses at the point focus. The 160 m^2 unit of ARUN160 was installed at Mahanand Dairy, Latur, India, for pasteurization of 20,000 L of milk daily. Fluids like water, steam, or thermic fluid can be used in the receiver with a delivery temperature of up to 200°C. Thermal storage can be used to store the thermal energy which can be supplied during the evening and night. This system can be used for providing process heat to chemical processing plants and a wide range of industries using boilers or heaters like sugar mills, textile mills, agro and food processing industries, vegetable oil mills, milk processing, drying of horticultural, food and fruits products, as well as units using vapor absorption refrigeration system for space cooling. This system can also be suitable for hospitals and hotels for providing hot water and steam [18].

5.2.6 Power tower or central receiver system

In power towers or central receiver systems, mirrored heliostats concentrate sunlight on a receiver placed at the top of a tower, and it is designed to reduce heliostat shadowing and increase optical efficiency. This type of point focus system can achieve high optical concentrations as well as high receiver temperatures and improved power cycle efficiency relative to a line-focus system can be obtained. A solar central receiver system consists of the following subsystems:

- (i) ***Collector system*** or heliostat field, consisting of several thousand planar mirrors called heliostats, individually tracking the sun with a two-axis system. The mirrors can be placed all around the tower, or only on the north side of the tower depending on the latitude.
- (ii) ***Solar receiver***, where the concentrated solar energy is absorbed. In the receiver, the working fluid or the HTF is heated with the help of concentrated solar energy. The fluids used in the receiver may be either molten salt of nitrates ($NaNO_3$ and KNO_3) or water. Other fluids, such as air, sodium, or helium, can also be used. A schematic diagram of a power tower with heliostats and solar receiver is shown in Fig. 5.9. The solar receiver may be of various types such as tubular, volumetric receiver, etc.
- (iii) ***Heat storage system***, which is used to provide energy to the working fluid when in sufficient sunlight is available during the nighttime or on acloudy day.
- (iv) ***Power block*** consisting of a steam generator and alternator.

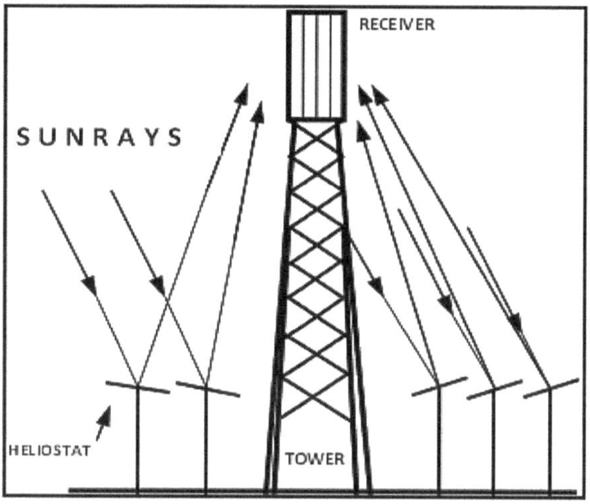

Figure 5.9 Schematic diagram of a solar central receiver with heliostats.

The sunlight is concentrated between 300 and 1500 times on the receiver due to which very high temperatures of the working fluid, ranging between 800 and 2000°C, can be reached. Solar tower power plants can thus reach higher power levels compared to other concentrated technologies, allowing for them to be best suited in utility scale applications ranging from 30 to 400 MWe or more. Ivanpah Solar Electric Generating System, situated in Mojave Desert, California, is currently the world's largest solar thermal power station, with a gross capacity of 392 MW. This facility deploys 173,500 heliostats, with two mirrors each, which focus solar radiation on boilers mounted on three solar power towers [19]. Some of the largest solar power tower systems are listed in Table 5.4 [15].

Table 5.4 Some of the largest solar power tower plants [15].

Name of power station	Location and country	Electrical capacity (MW)	Technology type
Ivanpah Solar Power Facility	County of San Bernardino, US State of California	392	Solar central receiver
Megalim Power Station (Negev Energy)	Ashalim, Israel	121	Solar central receiver
Cerro Dominador Solar Thermal Plant (Atacama 1)	María Elena, Antofagasta, Chile	110	Solar central receiver
Shouhang Dunhuang	Dunhuang (Gansu Province)	100	Solar central receiver

5.3 Thermal storage

Of the various renewable energy options available for power generation, concentrated solar thermal power generation is becoming a very attractive renewable energy production system, as it has better potential for dispatchability. This dispatchability is inevitably linked with an efficient and cost-effective thermal storage system. Intermittency of the solar resource for concentrating solar thermal power (CSP) plants can be overcome by thermal energy storage. When sufficient sunlight is not available, storage can utilize the power block, reducing the levelized cost of energy. Thus, of all the components of concentrating solar power, thermal storage is a key one [20]. However, it is also one of the less developed areas and only a few plants in the world have tested high-temperature thermal energy storage systems. When designing a thermal energy storage system for a concentrating solar power plant, the following points need to be considered:

(i) Cost of the system
(ii) Operating range of temperatures
(iii) The ease with which the system can be integrated into the power plant and
(iv) Maximum energy to be stored which is determined by the maximum load to be handled.

The three basic operations that are performed in a thermal energy storage system are shown in Fig. 5.10.

Various thermal energy storage media are available, which are, brief:

(i) ***Sensible heat storage***: Here, the temperature of the media varies and there is no phase transformation during the process. The amount of heat stored (Q) is given by the following expression:

$$Q = \int_{T_L}^{T_H} mC_p dT \tag{5.3}$$

where T_L and T_H are the initial and final temperatures, respectively, in K; m is the mass of heat storage medium in kg; and C_p is specific heat in J/kg-K or kJ/kg-K. Therefore, the amount of energy stored depends on: mass of storage media, value of specific heat, and the temperature difference. Sensible heat storage can be classified as solid and liquid media storage. Examples of solid media storage are rocks, sand, concrete, metals, etc., whereas liquid media storages are oil-based liquids, water in low-temperature range (normally within 25–90°C), molten salts, nanofluids, etc. Most of the currently installed thermal energy storage systems in solar thermal electric plants use sensible heat storage. Of these various options available, both oil and molten salt systems have been found to be technically feasible, but the use of thermal oil is restricted due to its flammability, high cost, and temperature limitations (400°C) due to which it is not possible to achieve high

Figure 5.10 Basic operations of a thermal energy storage system.

thermodynamic efficiency of the power block [21]. On the other hand, certain advantages of molten salt used in solar power tower systems are: (i) it is liquid at atmosphere pressure; (ii) it provides an efficient and less costly medium; (iii) it is nontoxic, nonflammable, and chemically inert; and (iv) its operating temperature matches with the high-pressure and high-temperature requirements of steam turbines.

(ii) ***Latent heat storage*:** The principle of latent heat storage is to store thermal energy, while the phase transformation of materials occurs (for example, solid to liquid state) at constant temperature. These materials are called phase change materials (PCMs) and molten salt is a common medium used for latent heat storage in solar thermal systems. The main advantage of latent heat storage as compared to other methods is that it has higher heat capacity, thus allowing for a smaller thermal storage unit. The phase changing of the PCM may be solid–solid, liquid–gas, or solid–liquid. Various types of PCMs are shown in Fig. 5.11. The common PCMs with their temperature ranges and latent heat of fusion are listed in Table 5.5 [22–25]. Of the various latent heat storage options, eutectics, more precisely a $NaNO_3$ and KNO_3 mixture, are widely used. $NaNO_3$ has a melting point of 307°C and KNO_3 of 337°C. However, when 46 mol% $NaNO_3$ and 54 mol% KNO_3 are mixed, the mixture has a eutectic point of 222°C, which is lower than the individual melting point of salts.

Certain advantages of latent heat storage are (i) high latent heat capacity, (ii) high thermal conductivity, and (iii) nonflammable.

(iii) ***Thermo-chemical heat storage*:** The principle of thermo-chemical heat storage (TCS) is that during an endothermic chemical reaction, while new product(s) is/are formed, energy is stored and the energy is released during an exothermic chemical reaction. A higher storage capacity than LHS is possible with this method depending on the reactants.

Several thermal energy storage technologies are available such as the following:

(i) Two-tank direct storage system
(ii) Two-tank indirect storage system and
(iii) Single-tank thermocline system.

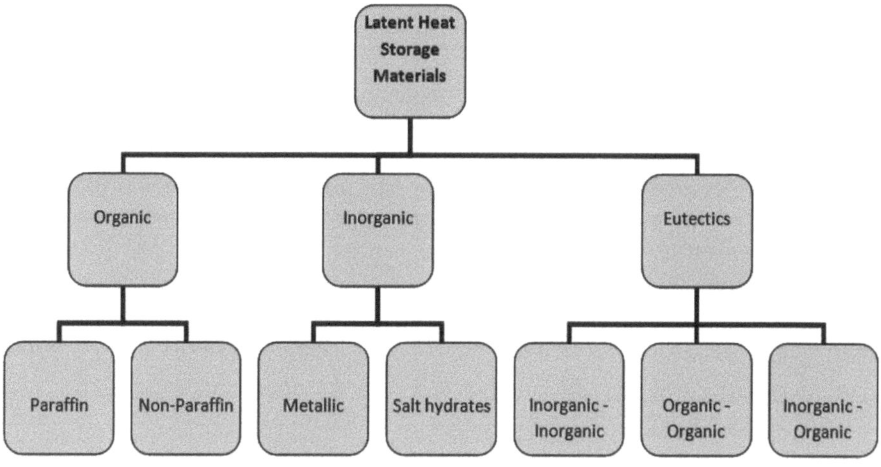

Figure 5.11 Various types of PCMs.

Table 5.5 PCMs with their temperature ranges and latent heat of fusion [22–25].

Temperature range (°C)	Material	Transition temperature (°C)	Heat of fusion or latent heat (kJ/kg)
0–100	Ice	0	335
	n-Octane	27.7	243.5
	Paraffin Wax	32	251
	Na-acetate trihydrate	58	250
100–400	Erythritol	118	340
	$LiNO_3$–$NaNO_3$	195	252
	$NaNO_3$	306	172
	KNO_3	333	266
	KOH	380	150
400–800	50LiOH/50LiF	427	512
	$KClO_4$	527	1253
	LiH	699	2678
800–1500	LiF	868	932
	NaF	993	750
	MgF_2	1271	936
	Si	1415	1654

5.4 Power block

The power block is the "heart" of the CSP plant. Here, the thermal energy delivered to the working fluid (i.e., generally air or water) either from storage or from the solar field is converted into electrical energy via a turbine and alternator. In this section, a closer look is given to how the thermal energy is converted into electrical energy. For CSP plants, the main thermodynamic cycles used to produce power are Rankine, Brayton, and combined cycles.

(i) ***Rankine cycle*:** Fig. 5.12 shows a schematic model of a Rankine cycle-based concentrated solar thermal power plant (CSTPP) and Fig. 5.13 presents its corresponding temperature (T)–entropy (S) diagram (with no reheat). While the dotted lines represent the real processes, the continuous lines show the ideal processes. The four main processes are depicted below:

***Process A*:** Irreversible compression in the feed water pump, pump power being P_P

***Process B*:** Steam generation in the steam generator (economizer, evaporator, and superheater). Required heat is Q_{added}

***Process C*:** Steam turbine power (P_T)

***Process D*:** Heat rejected ($Q_{rej.}$) in the condenser.

Therefore, the thermal efficiency η_R of the Rankine cycle is given by the following:

$$\eta_R = \frac{Net\ Output\ Power}{Heat\ Added} = \frac{(P_T - P_P)}{Q_{added}} \tag{5.4}$$

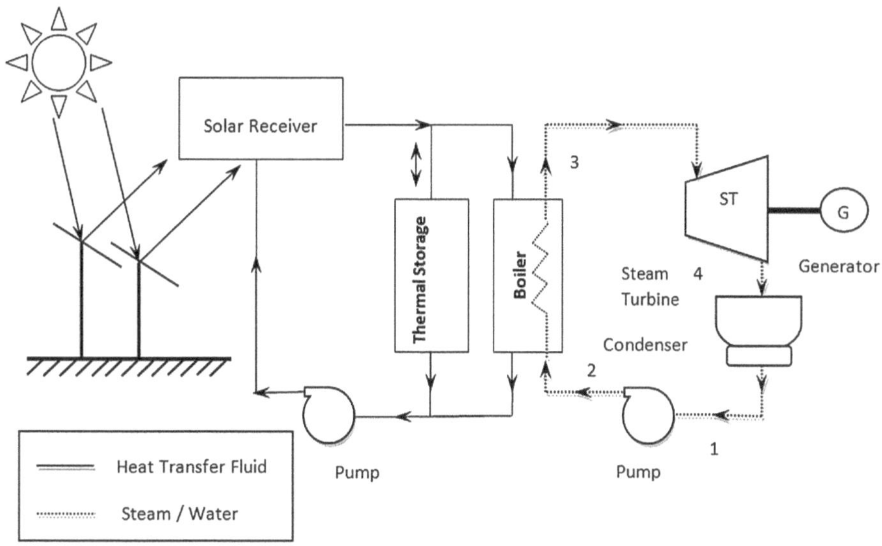

Figure 5.12 Rankine cycle-based concentrated solar thermal power plant.

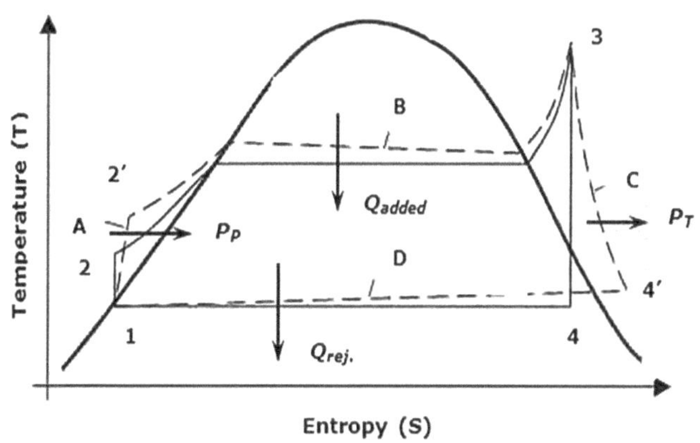

Figure 5.13 Temperature–entropy diagram of the Rankine cycle.

(ii) ***Brayton cycle*:** The actual gas turbine Brayton cycle is different from the ideal one due to irreversibilities that are involved in the actual cycle. Therefore, in an actual gas turbine cycle, the compressor consumes more work and the turbine produces less work than that of the ideal Brayton cycle. The actual power input to the compressor and the actual power output from the turbine can be obtained using the adiabatic efficiencies of the compressor and turbine. Fig. 5.14 shows a schematic diagram of a Brayton cycle-based concentrated solar thermal system and its corresponding temperature (T)–entropy (S) diagram is shown in Fig. 5.15.

***Process 1–2*:** Compression in compressor
***Process 2–3*:** Heat input to the cycle (Q_{in})
***Processes 3–4*:** Gas turbine work

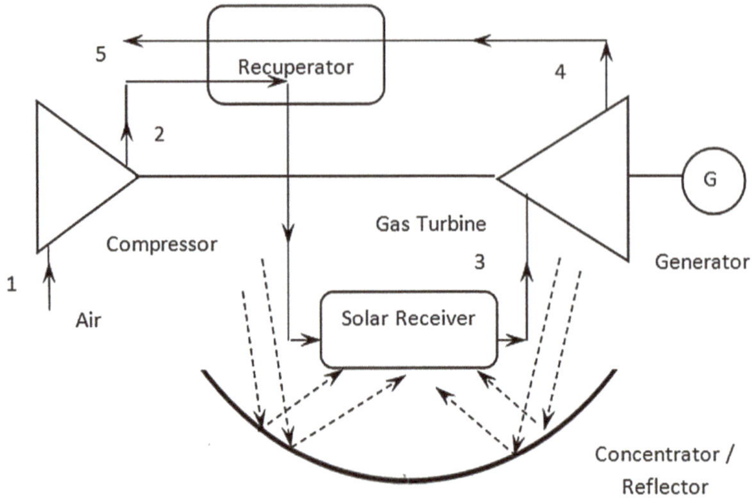

Figure 5.14 Brayton cycle-based concentrated solar thermal system.

Figure 5.15 Temperature (T)–entropy (S) diagram of the Brayton cycle.

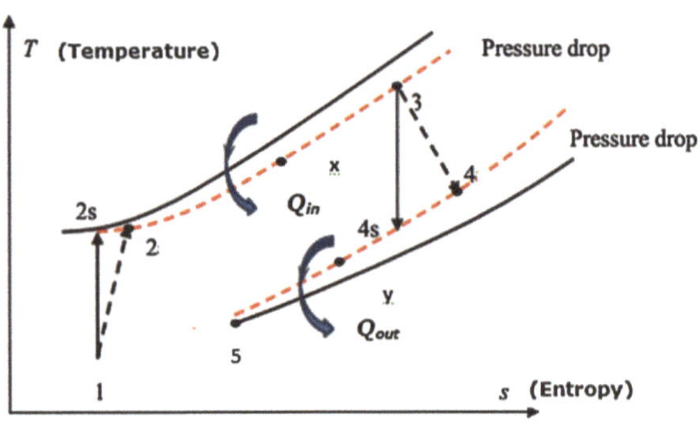

Process 4–1: Heat rejection (Q_{out}).

If r is the pressure ratio and k is the ratio of specific heat of the working fluid of the Brayton cycle, then the thermal efficiency of the Brayton cycle can be expressed as follows:

$$\eta_{th,Brayton} = 1 - \frac{1}{r^{(k-1)/k}} \tag{5.5}$$

(iii) ***Combined cycle***: A schematic diagram of a concentrated solar thermal power system with a combined cycle is shown in Fig. 5.16. Topping cycle is a Brayton cycle where air is the working fluid. Compressed air passes through the solar receiver to obtain the necessary heat energy. Exhaust air from a gas turbine passes through the HRSG to produce superheated steam which drives the steam turbine. Therefore, combined power is obtained from gas as well as steam turbine.

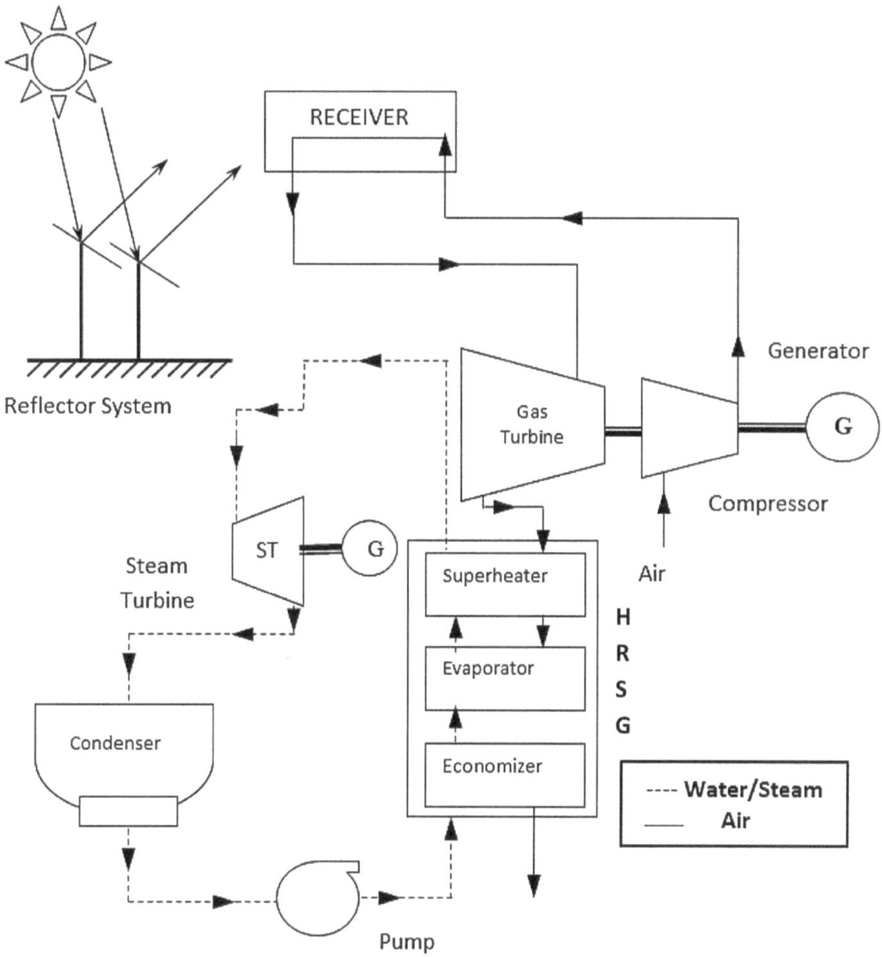

Figure 5.16 A concentrated solar thermal power system with combined cycle.

6. Municipal solid waste as renewable energy

MSW, often called garbage or trash, is the waste primarily produced by households, but it also includes some wastes from industries which are similar in nature to the wastes from households. It contains bioproducts such as cardboard, paper, grasses, food wastes, wood, leaves, and leather/plastic products. These wastes are generally deposited in municipal landfill sites. MSW can be a liability to a country if not handled properly. In landfill, MSW decomposes and emits methane, which is a greenhouse gas that is 23 times more dangerous than carbon dioxide from the point of view of environmental pollution. Another part, i.e., leachate, also causes serious environmental

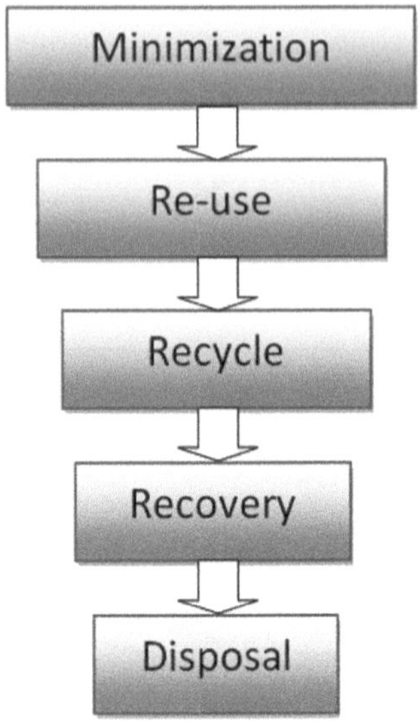

Figure 5.17 Waste hierarchy of MSW.

pollution in ground water and air. Many developed countries have adopted recovery and recycle options from the principle of waste hierarchy, as shown in Fig. 5.17, to reduce the final disposal of MSW.

Still, significant quantities of MSW are disposed of in landfills after recycling even in countries with highly developed recycling infrastructure, and utilization of MSW has become a challenge to many societies. Hence, energy recovery from MSW is an environmentally justified and economically viable option. It has been reported by the World Bank that while 1.3 billion tonnes of waste is generated currently per year globally, it will reach 2.2 billion tonnes per year by 2025 [26]. This has directed policy makers to ponder over and implement a new waste management system.

The energy recovery option from municipal solid waste is a promising option which can reduce the mass and volume of wastes considerably. It has been observed that about 80% of the carbon in MSW is biomass derived, and thus it can be considered as a renewable source [27]. Further, MSW has been listed as a renewable energy source by the United States Environmental Protection Agency. Waste to energy technologies (WTE-T) play a vital role in the sustainable development and management of MSW worldwide. These technologies involve the recovery of heat and electricity from MSW, especially from nonrecyclable wastes [28,29]. This technology is rapidly growing all over the world because [29–33]

(i) It reduces the demand for landfills
(ii) Dependence on fossil fuels is reduced
(iii) Greenhouse gas (GHG) emissions are reduced, and
(iv) It has a very good impact on economic growth.

6.1 Types of municipal solid waste

Various types of municipal solid waste are shown in Fig. 5.18.

Of the various options available in a thermal conversion process, incineration is widely used in developed as well as developing countries as a waste management technology for the production of steam and electricity [34,35]. Though this process has several advantages, the following major disadvantages exist in this process:

(i) It is an expensive process.
(ii) It produces acid gases, carcinogen dioxin, NO_x, particulates, and heavy metals. These pollute the atmosphere and dioxin is carcinogenic.
(iii) The process has long-term effects on public health, as observed in research. It has been observed that long-term health issues such as cancer, birth defects, reproductive dysfunction, and neurological problems occur in communities where incineration plants are located.
(iv) The ash which remains as residue in the process contains heavy metals which require further treatment to avoid further health and environmental issues.

Gasification, which is also a thermal conversion technology, is considered to be a potential candidate for converting MSW to energy [36], as it produces less emissions in comparison to incineration [37]. In this thermal conversion process, organic

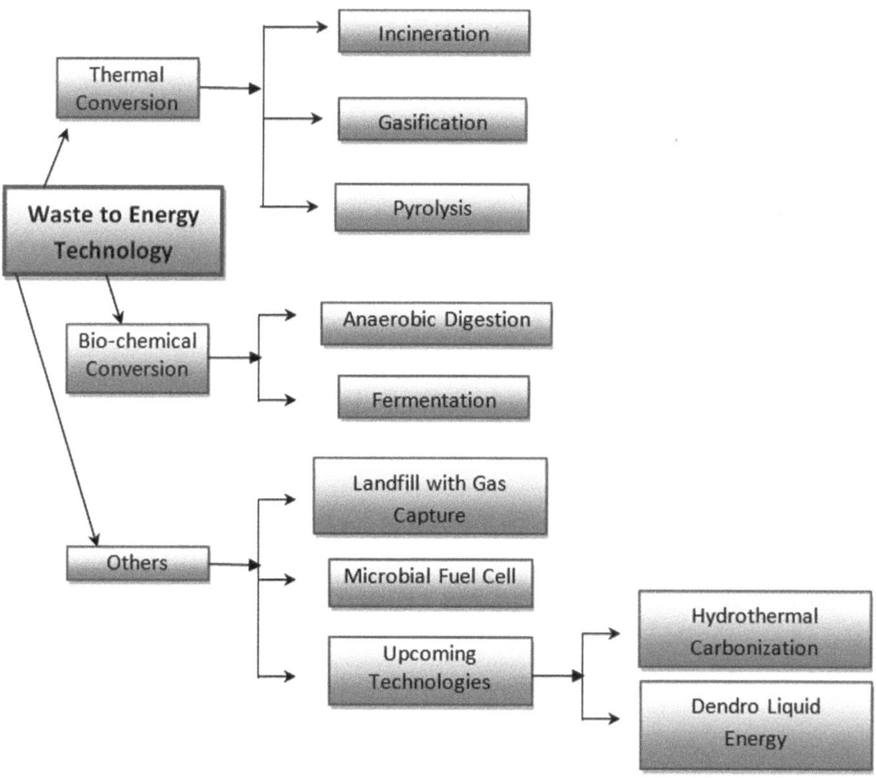

Figure 5.18 Types of municipal solid waste.

compounds are converted into syngas in an atmosphere of oxygen in substoichiometric ratio at high temperature. The main product of gasification is syngas, which can be further combusted to supply heat to produce steam or the products of combustion can be used to run a gas turbine to produce electricity through an alternator. Fig. 5.19 shows the different categories of biomass gasification.

6.2 Gasification process of municipal solid waste

The MSW gasification process is illustrated through the schematic diagram in Fig. 5.20. There are four basic operations in gasification:

(i) ***Preprocessing*:** This involves the washing, drying, sieving, and thermal pretreatment of raw MSW.
(ii) ***MSW gasification and generation of syngas*:** After preprocessing, MSW is fed to the gasifier to produce syngas.
(iii) ***Syngas clean-up*:** In this process, syngas is cleaned.
(iv) ***Gas utilization*:** The syngas is used in a power block to generate electricity.

6.3 Power block

In harnessing electricity or energy from MSW gasification, the power block is a very important unit. Electricity can be obtained from syngas mainly from the following ways:

(i) Gas turbine cycle,
(ii) Steam turbine cycle, and
(iii) Combined cycle.

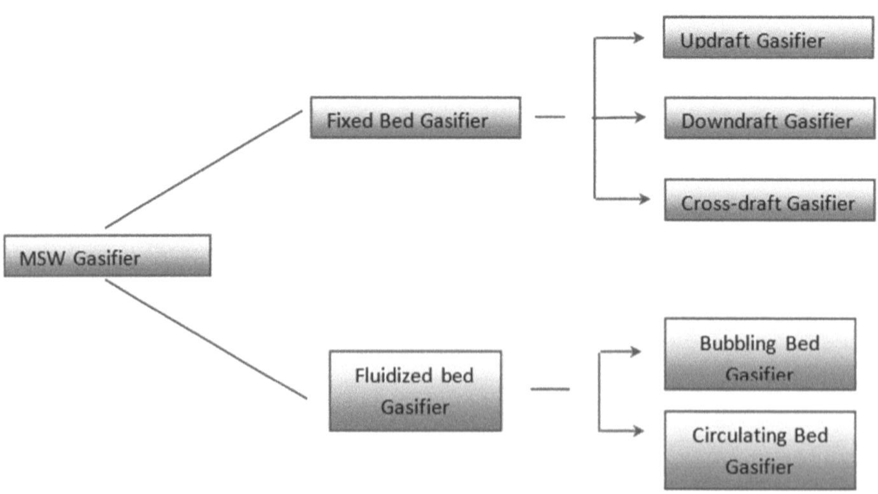

Figure 5.19 Different types of MSW gasifier.

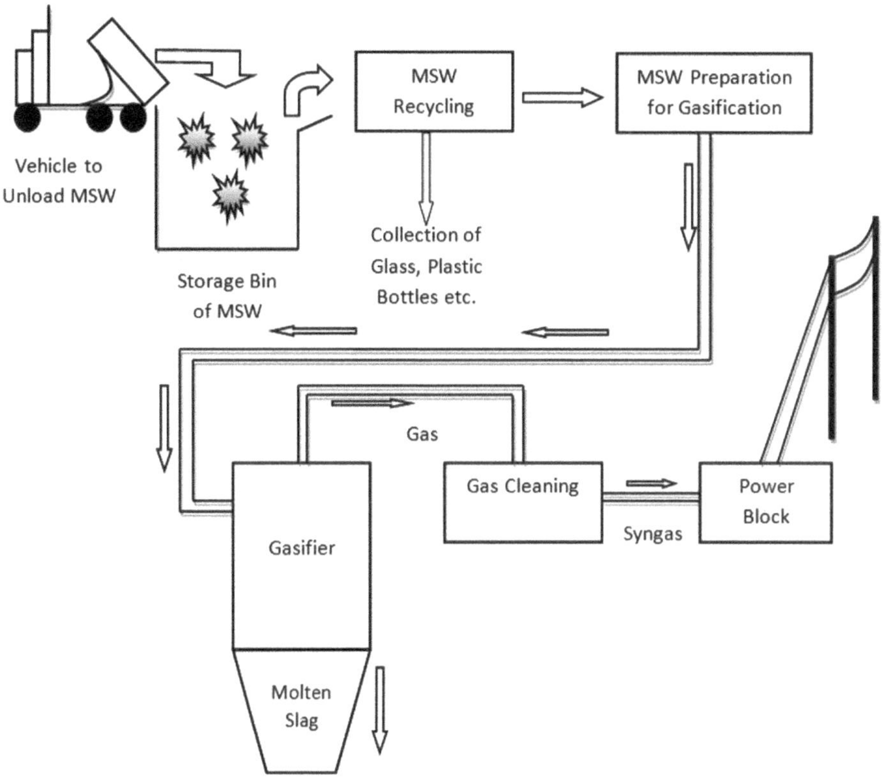

Figure 5.20 MSW gasification process.

6.4 Gas turbine cycle

In a conventional gas turbine (GT) cycle or directly fired gas turbine cycle (DFGT), hot and compressed air after passing through the compressor and combustion chamber enters the gas turbine to produce power. In the combustion chamber, syngas from the gasifier is combusted with the help of air. The gas turbine is coupled with a generator to produce electricity. A model of a DFGT cycle is shown in Fig. 5.21.

However, in an indirectly heated/externally fired gas turbine (EFGT) cycle, the hot combustion product does not come in direct contact with the turbine blades. In this process, the heat is provided by placing the combustor in the hot exhaust air stream from the turbine and a high-temperature heat exchanger (HTHE) is used to separate the combustion products from the gas turbine blades and to preheat the gas turbine cycle working fluid, generally air. Fig. 5.22 shows a schematic diagram of an EFGT cycle.

6.5 Steam turbine cycle

A steam turbine model with syngas produced from a gasifier is shown in Fig. 5.23. The syngas with the help of air is combusted in a combustion chamber. The gas produced is

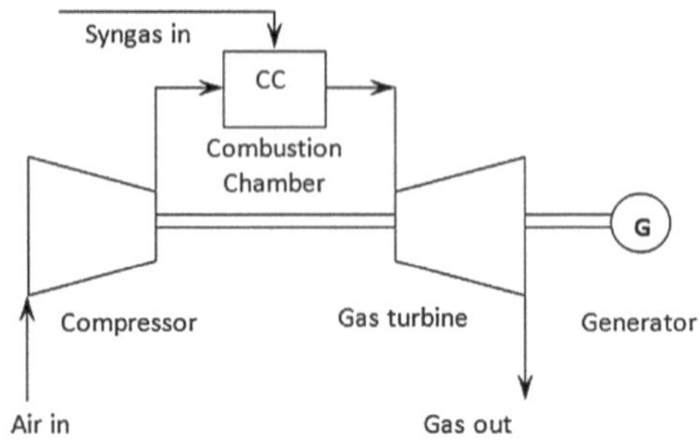

Figure 5.21 Configuration for energy recovery of syngas from a DFGT cycle.

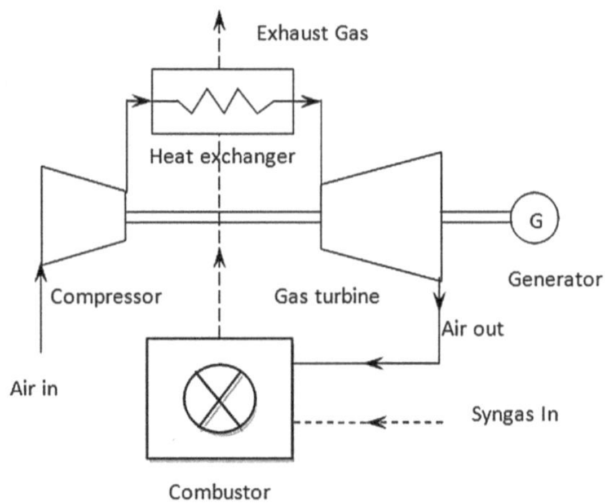

Figure 5.22 Configuration for energy recovery of syngas from an EFGT cycle.

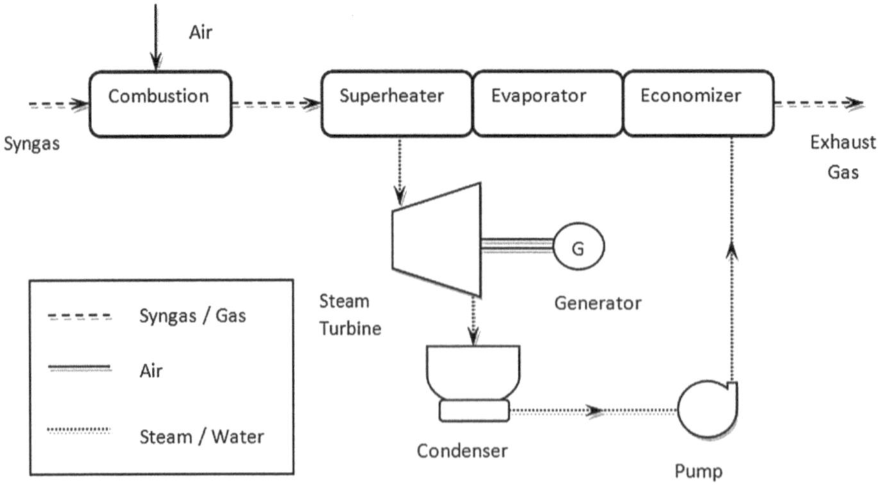

Figure 5.23 Configuration for energy recovery by a simple steam cycle with syngas.

used to produce superheated steam in an HRSG/boiler which runs a steam turbine. The steam turbine is coupled with a generator (*G*) to generate electricity.

6.6 Combined cycle

There may be various configurations of MSW combined cycle power plant. A schematic diagram of an externally fired combustion chamber (EFCC) gasifier combined cycle plant is shown in Fig. 5.24. MSW, after proper cleaning and segregation, enters the gasifier (*G*) and is gasified in the presence of substoichiometric conditions of atmospheric air supply.

After the gasification process, syngas is generated and enters the combustion chamber (CC). The syngas mixture is combusted in the presence of recirculated gas turbine (GT) exhaust and thus flue gas is generated. Flue gas then enters the shell side of a heat exchanger (HX), which in turn heats up the working medium of the topping gas turbine cycle. The combustion chamber combined with the heat exchanger constitutes a combustor-heat exchanger duplex (CHX) unit. Atmospheric air enters the compressor (*C*) and after being compressed enters the tube side of the heat exchanger (HX) and is heated. The hot and compressed air then enters the gas turbine and gets expanded. The gas turbine exhaust is recirculated to the combustion chamber for better combustion of syngas. The compressor is driven by the gas turbine and the excess shaft power, after driving the compressor, drives an electric generator to produce electricity.

Exhaust from the CHX unit enters a heat recovery steam generator (HRSG) to run a bottoming Rankine cycle. The HRSG unit consists of three subunits, i.e., a superheater, evaporator, and economizer. The Rankine cycle is constituted of the HRSG, steam turbine (ST), condenser, and a pump. Finally, the flue gas from the HRSG is exhausted to the atmosphere through a stack.

6.6.1 Performance parameters

Net power output from a combined cycle is the sum of power outputs from a gas turbine and a steam turbine and calculated as follows:

$$W_{\mathrm{CC}} = W_{\mathrm{GT}} + W_{\mathrm{ST}} - W_C - W_P \tag{5.6}$$

where W_{CC}, W_{GT}, and W_{ST} are the power obtained from combined cycle, gas turbine, and steam turbine, respectively, and W_C and W_P are the power required to run the compressor and pump, respectively.

The thermal efficiency or the first law efficiency of the MSW combined cycle is expressed as follows:

$$\eta_{\mathrm{CC}} = \frac{W_{\mathrm{CC}}}{m.\mathrm{LHV}} \tag{5.7}$$

where *m* and LHV are the mass flow rate and lower heat value of municipal solid waste, respectively.

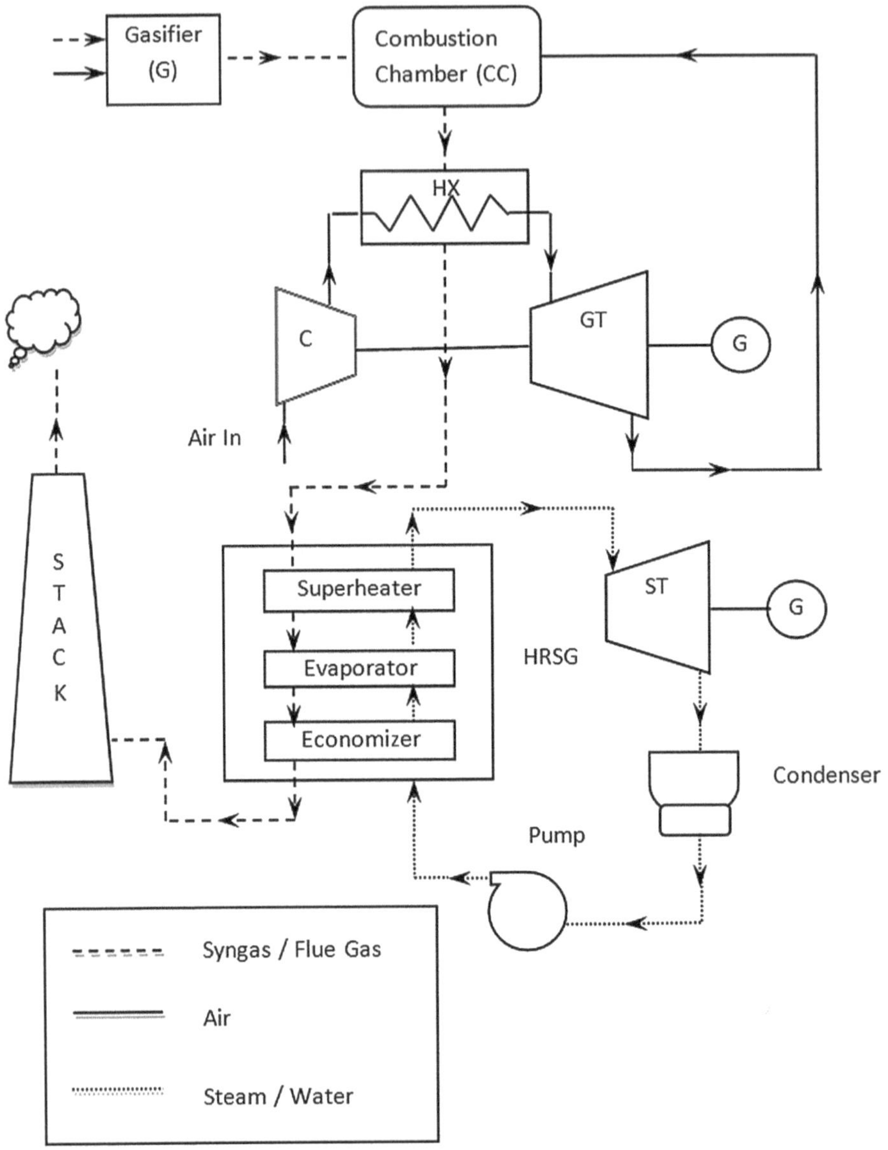

Figure 5.24 Schematic diagram of an externally fired combustion chamber (EFCC) gasifier combined cycle plant.

Electrical specific MSW consumption (ESMC) is defined as the mass of municipal solid waste required to produce 1 unit kW of electricity and is calculated as follows:

$$\mathrm{ESMC} = \frac{m * 3600}{W_{CC}} \tag{5.8}$$

The exergetic efficiency or the second law efficiency of any component is given by the following:

$$n_{\text{exergetic}} = \frac{ex_{\text{product}}}{ex_{\text{input}}} \tag{5.9}$$

7. Conclusion

It is a fact that the lives of all fauna and flora in the world are at risk mainly due to the greenhouse gas emissions and this risk will increase daily if there is no balance between ecology and the environment, for which all countries are worried and trying to shift to sustainable energy sources. It is necessary to make a sincere effort to emit "zero" GHGs by 2050, otherwise it will be difficult for the long-term prospects of civilization. To meet net zero by 2050, available clean and efficient energy technologies should be deployed in this decade along with the implementation of strong policies by governments for the use of sustainable energy sources.

In this chapter, a detailed discussion has been conducted regarding two promising options for renewable and/or sustainable energy sources: solar energy and energy harnessed from MSW. Various models of these two sources along with their performance parameters have been discussed. Though solar PV is a mature technology, CSP is also getting its place to meet the energy demand for small-, medium-, and utility-scale applications. For continuous power supply from CSP systems, heat storage should be incorporated, which needs more research to increase its efficiency.

Energy from MSW is another promising source of sustainable energy and it should be treated as a resource to be exploited rather than a waste requiring disposal. Though there are various methods of obtaining electricity from the feedstock of municipal solid wastes, gasification is a promising option which has been discussed in detail along with different energy models.

References

[1] Paris Agreement, United Nations, 2015 available at, https://unfccc.int/sites/default/files/english_paris_agreement.pdf. (Accessed 24 May 2021).
[2] World Energy Scenarios Composing Energy Futures to 2050, World Energy Council, 2013. https://www.worldenergy.org/assets/downloads/World-Energy-Scenarios_Composing-energy-futures-to-2050_Full-report1.pdf. (Accessed 24 May 2021).
[3] S. Kumar, R. Madlener, CO_2 emission reduction potential assessment using renewable energy in India, Energy 97 (2016) 273–282.
[4] G. Schmid, The development of renewable energy power in India: which policies have been effective? Energy Policy 45 (2012) 317–326.
[5] Statistical Review of World Energy, 2020 available at, https://www.bp.com/content/dam/bp/business-sites/en/global/corporate/pdfs/energy-economics/statistical-review/bp-stats-review-2020-full-report.pdf. (Accessed 24 May 2021).

[6] C.A.S. Hall, J.G. Lambert, S.B. Balogh, EROI of different fuels and the implications for society, Energy Policy 64 (2014) 141−152.

[7] 2010 Renewable Energy Data Book and 2018 Renewable Energy Data Book, U. S. Department of Energy, https://www.nrel.gov/docs/fy12osti/51680.pdf and https://www.nrel.gov/docs/fy20osti/75284.pdf. Accessed on 30. May 2021.

[8] Renewable Energy and Jobs Annual Review 2020, International Renewable Energy Agency (IRENA) available at: https://www.irena.org/publications/2020/Sep/Renewable-Energy-and-Jobs-Annual-Review-2020. Accessed on 01. June 2021.

[9] S.P. Sukhatme, J.K. Nayak, Solar Energy Principles of Thermal Collection and Storage, Tata McGraw-Hill Publishing Company Limited, New Delhi, 1996, p. 23.

[10] Benefits of Rooftop Solar Panels and Factors that Further Aid Their Installation in India, 2019. https://economictimes.indiatimes.com/small-biz/productline/power-generation/benefits-of-rooftop-solar-panels-and-factors-that-further-aid-their-installation-in-india/articleshow/69233619.cms?from=mdr. (Accessed 1 June 2021).

[11] W.A. Thomas, Archimedes through the looking glass, Classical World 68 (5) (1975) 305−308.

[12] M. Brenna, F. Foiadelli, M. Roscia, D. Zaninelli, Evaluation of solar collector plant to contribute climate change mitigation, in: Proceedings of the IEEE International Conference on Sustainable Energy Technologies, ICSET, Singapore, 2008, pp. 198−202.

[13] N. Tanak, Technology Roadmap: Concentrating Solar Power, International Energy Agency, 2010 available at, https://iea.blob.core.windows.net/assets/663fabad-397e-4518-802f-7f1c94bc2076/csp_roadmap.pdf. (Accessed 12 June 2021).

[14] K.M. Powell, F.T. Edgar, Modeling and control of a solar thermal power plant with thermal energy storage, Chemical Engineering Science 71 (2012) 138−145.

[15] List of solar thermal power stations, available at: https://en.wikipedia.org/wiki/List_of_solar_thermal_power_stations Accessed on 16. June 2021.

[16] R.M. Thomas, A Report on Solar-Electric Dish Stirling System Development, Report Number: SAND-97-2924C CONF-980211, 1997.

[17] D. Mills, Advances in solar thermal electricity technology, Solar Energy 76 (2004) 19−31.

[18] Solar Concentrator for Industrial Process Heat, available at: http://www.ircc.iitb.ac.in/IRCC-Webpage/PDF/update/archives/Issue1_2006/solar-concentrator.html. Accessed on 23. June 2021.

[19] Ivanpah Solar Power Facility, https://en.wikipedia.org/wiki/Ivanpah_Solar_Power_Facility. Accessed on 23. June 2021.

[20] U. Hermann, D. Kearney, Survey of thermal storage for parabolic trough power plants, Journal of Solar Energy Engineering, Transactions of the ASME 124 (2002) 8.

[21] S.M. Flueckiger, Z. Yang, S.V. Garimella, Review of molten-salt thermocline tank modeling for solar thermal energy storage, Heat Transfer Engineering 34 (10) (2013) 787−800.

[22] I. Sarbu, C. Sebarchievici, A comprehensive review of thermal energy storage, Sustainability 10 (2018) 191.

[23] G. Alva, L. Liu, X. Huang, G. Fang, Thermal energy storage materials and systems for solar energy applications, Renewable and Sustainable Energy Reviews 68 (2017) 693−706.

[24] Z. Khan, Z. Khan, A. Ghafoor, A review of performance enhancement of PCM based latent heat storage system within the context of materials, thermal stability and compatibility, Energy Conversion and Management 115 (2016) 132−158.

[25] S.M. Hasnain, Review on sustainable thermal energy storage technologies, Part I: heat storage materials and techniques, Energy Conversion and Management 39 (11) (1998) 1127−1138.

[26] D. Hoornweg, P. Bhada-Tata, What a Waste: A Global Review of Solid Waste Management, 2012.
[27] *A position paper prepared by IEA Bioenergy*, Municipal Solid Waste and its Role in Sustainability, IEA Bionergy, 2003.
[28] C. Ryu, D. Shin, C. Ryu, D. Shin, Combined heat and power from municipal solid waste: current status and issues in South Korea, Energies 6 (2012) 45−57.
[29] D. Zhang, G. Huang, Y. Xu, Q. Gong, Waste-to-Energy in China: key challenges and opportunities, Energies 8 (2015) 14182−14196.
[30] C. Ofori-Boateng, K.T. Lee, M. Mensah, The prospects of electricity generation from municipal solid waste (MSW) in Ghana: a better waste management option, Fuel Processing Technology 110 (2013) 94−102.
[31] A.U. Zaman, S. Lehmann, The zero waste index: a performance measurement tool for waste management systems in a "zero waste city", Journal of Cleaner Production 50 (2013) 123−132.
[32] M. Khan, H. Tanveer, Production of thermoelectric power from solid waste of Urban Lahore, in: Proceedings of the 2011 International Conference & Utility Exhibition on Power and Energy Systems: Issues and Prospects for Asia, ICUE, 2012, https://doi.org/10.1109/ICUEPES.2011.6497736.
[33] C. Psomopoulos, I. Venetis, N.J. Themelis, T.E.I. Piraeus, The impact from the implementation of "waste to energy" to the economy. A macroeconomic approach for the trade balance of Greece, Fresenius Environmental Bulletin 23 (2014) 2735−2741.
[34] L. Zheng, J. Song, C. Li, Y. Gao, P. Geng, B. Qu, L. Lin, Preferential policies promote municipal solid waste (MSW) to energy in China: current status and prospects, Renewable and Sustainable Energy Reviews 36 (2014) 135−148.
[35] A. Beylot, J. Villeneuve, Environmental impacts of residual municipal solid waste incineration: a comparison of 110 French incinerators using a life cycle approach, Waste Management 33 (2013) 2781−2788.
[36] H.A. Arafat, K. Jijakli, Modeling and comparative assessment of municipal solid waste gasification for energy production, Waste Management 33 (2013) 1704−1713.
[37] A. Kumar, S.R. Samadder, A review on technological options of waste to energy for effective management of municipal solid waste, Waste Management 69 (2017) 407−422.

RF energy harvesting

6

Geetha Palani[1], *A. Arputhalatha*[2], *Karthik Kannan*[3], *Poobana Dharmalingam*[4] *and S. Esther Gandhimathi*[5]
[1]Department of Physics, Faculty of Science and Technology, J.N.N. Institute of Engineering, Chennai, Tamil Nadu, India; [2]Department of Physics, Arizona State University, Tempe, AZ, United States; [3]BrainPool Program Postdoctoral Fellow, School of Advanced Materials Science and Engineering, Kumoh National Institute of Technology (KIT), Gyeongbuk, Republic of Korea; [4]School of Biosciences and Technology, VIT University, Vellore, Tamil Nadu, India; [5]Department of Electronics and Communication Engineering, Dhanalakshmi College of Engineering, Chennai, Tamil Nadu, India

1. Introduction

An unprecedented development in the field of wireless power transfer (WPT) has been witnessed in the last decade. WPT has a long history, dating back to the 18th century [1,2], which began with microwave power transmission. Heinrich Hertz conducted experiments in the 1880s to prove Maxwell's theory. Following that, during World War II, Brown conducted WPT tests by utilizing microwave technology of high efficiency based on wireless communications and radar remote sensing [3]. R. George developed the first rectenna in 1963, which consisted of each coupled to a bridge rectifier made up of an array of 28 half-wave dipole-receiving antennas and four 1N82G point-contact semiconductor diodes [4]. When operating the duration of two and 3 GHz with output powers of 7 and 4 W DC, this rectenna had efficiencies of 40% and 50%, respectively. Because of the diodes' low power-handling capability, DC outputs were quite low, making this rectenna unsuitable for powering up a helicopter rotor as an ambitious goal. An impedance-matching network and the problem of most of the power being reflected at the antenna–diode interface, because of the rectenna's low efficiency, it consisting of a plain array of rods was used to improve efficiency. Specifically, the matching network and the 4480 1N82G diodes were used to generate a substantially greater output power of 270 W; which is suitable for helicopter rotor powering. Following that, on July 1, 1964 [5], the newly created rectenna was used to successfully fly a microwave-powered helicopter.

Technology's recent achievements have led to the development of the Internet of Things (IoT), such as wearable electronics, 5G wireless systems, and so on; it needs sensors deployment worldwide at remote places it often positioned [6] communicating wirelessly with each other with the capacity. Because we must rely on battery sources, power is frequently a limiting factor. This results in a time-consuming disposal task and a large number of battery replacements, which pollute the environment. A green and more sustainable solution to this problem is offered by energy harvesting (EH).

Green Energy Systems. https://doi.org/10.1016/B978-0-323-95108-1.00007-0

This entails absorbing and converting ambient energy to electricity to charge sensors and batteries [7]. Among the ambient energies, there are various sources such as electromagnetic energy, mechanical energy, wind energy, solar energy, tidal energy, thermal energy, and so on. Energy-harvesting systems, which scavenge energy from ambient power wireless sensor devices to environment sources, have emerged as viable alternatives to batteries.

Fig. 6.1 depicts ambient energy sources, for example thermal energy, RF energy, vibration energy, and solar energy. Vibration energy harvesters use the piezoelectric effect to transform human or machine movement into electrical energy. However, the individual or machine cannot operate constantly and will need to rest. Thermal energy is converted into electrical energy by thermoelectric energy harvesting due to the thermal temperature difference between the two objects. This is only conceivable in the presence of a temperature gradient in the environment, which is uncommon in many applications [9]. The high cost and limited efficiency of thermoelectric energy harvesting are other disadvantages. Solar energy harvesting uses the photovoltaic (PV) effect to turn sunlight into electrical energy. Even though solar energy is the most abundant energy source, it is not a promising energy source for wireless sensors that are normally placed indoors and run day and night [10]. The power conversion efficiency (PCE) of a typical solar cell is approximately 20%. Radiofrequency energy harvesting is the method of converting RF energy into electrical energy that can be used. RF energy is emitted by wireless radio networks, television (TV) towers, and cellular towers, and can be found practically anywhere in the environment.

1.1 Motivation for wireless energy harvesting

One location in the Earth's atmosphere generates electrical power and another location receives it without any other supporting medium or wires of a wireless power

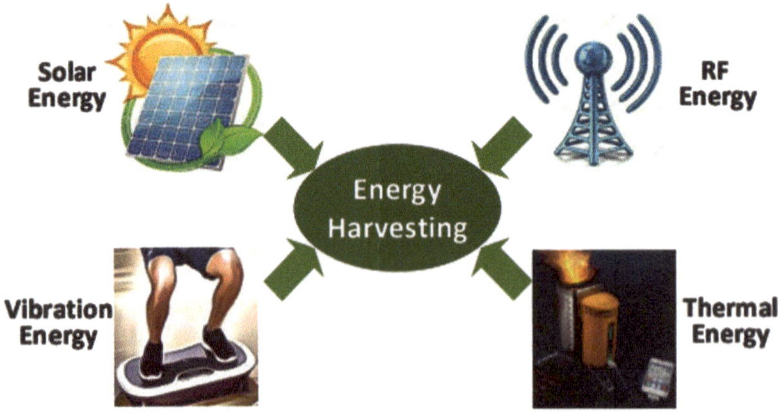

Figure 6.1 Different types of energy sources for energy harvesting.
Reprinted with permission from D. Khan, et al., A survey on RF energy harvesting system with high-efficiency RF-DC converters, Journal of Semiconductor Engineering 1 (1) (2020).

transmission system, is depicted as an early definition of energy harvesting [11]. A microwave-powered helicopter system launched RF power scavenging in free space in the late 1950s history. Later, energy scavenging or power harvesting was defined as a strategy in the external environment for extracting energy utilizing a variety of processes such as vibrational stimulation, thermoelectric conversion, pressure gradients, and solar energy conversion. This method has a lot of potential in low-power electrical devices for replacing small batteries and systems [12]. The construction of an RF energy-harvesting system as well as the parameters that influence the system's overall performance are depicted in Figs. 6.1 and 6.2.

For replacing or extending the life of batteries, RF wireless power harvesting (WPH) has great potential. The bulk of embedded equipment and low-power remote sensing devices are currently powered by batteries. However, batteries have a limited lifespan and must be replaced regularly. Equipment and devices may become self-sustaining in terms of the operating energy required by utilizing power-harvesting technology, resulting in an infinite functional lifespan. As a result, there will be no need for electricity maintenance [14].

Thermal energy, solar power, wind energy, kinetic energy, electromagnetic energy, and other sources of WPH are also available. Electromagnetic energy, for example, is abundant in space and could be collected indefinitely. Transmission, radio stations, digital multimedia, wireless internet, and satellite stations are all sources of electromagnetic waves. An RF power-harvesting system captures electromagnetic energy and converts it into a useful direct current (DC) voltage [15]. The rectifier circuit and antenna, where DC energy is converted to RF power or alternating current (AC), are the most important components of an RF power harvesting-system.

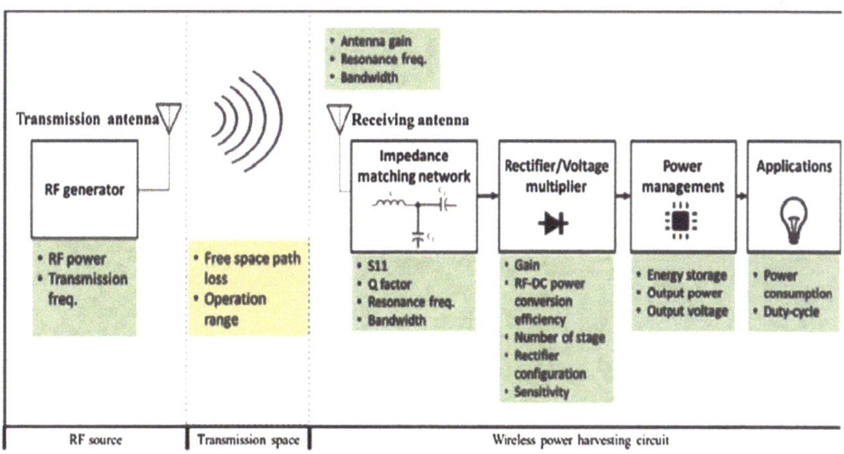

Figure 6.2 RF power-harvesting system conceptual block diagram.
Reprinted with permission from L.-G. Tran, et al., RF power harvesting: a review on designing methodologies and applications, Micro and Nano Systems Letters 5 (2017) 14.

Battery waste processing is a major issue. Most batteries are disposed of in landfills, polluting the ground and water. Avoiding the use of batteries is the most efficient way to reduce battery waste. WPH technology will assist in minimizing reliance on batteries, which will have a good effect on the environment in the long term [16]. Additionally, because electromagnetic energy is a clean energy source, the process of harvesting it produces no waste.

2. RF energy harvesting

RFE is of particular importance among the different energy-harvesting sources because of its omnipresence in the environment, regardless of the season of the year, day, time, and most significantly because it allows mobility. The difficulty here is in storing or harnessing this ambient energy in a small amount; it may be used as a backup for remote power with battery backup or a battery recharge power source [17,18]. This can be used for devices as direct power that does not require batteries, for example, wireless sensors. It can be scaled to many nodes without source code modification, as well as being inserted within walls. As a result, RFE outperforms other harvesting methods. Harvesting energy from these sources, on the other hand, presents certain difficulties; power is very low for ambient RFE because it is inversely proportional to the distance from the source, necessitating the use of high-gain antennas. To obtain high efficiency, radiofrequency to DC converter's power dissipation should be used with minimal. Because of barriers in the transmission line, RFE can be reflected and absorbed. For optimum power transfer, broadband impedance matching is also critical in the band at all required frequencies when using multiband radio frequency harvesting.

The following are several sources of ambient RF [19]:

- GSM band (0.85–0.90 GHz, 1.8–1.9 GHz);
- AM radio band (550–1605 kHz);
- TV band (41–950 MHz);
- WIFI band (2.45–5.8 GHz);
- 4G band (2–8 GHz);
- 3G band (1.8–2.5 GHz).
- FM radio band (87.5–108 MHz);
- Code Division Multiple Access (CDMA) band (450–2100 MHz);
- ISM band (2.4 GHz).

An RF energy-harvesting circuit, as discussed in this chapter, typically comprises an antenna, a rectifier, a voltage multiplier, an impedance matching circuit, load, and an energy storage device. Fig. 6.3 shows a block diagram of an energy harvester. The transmitted RF signal travels via the wireless transmission medium in an RF energy-harvesting system, by a single antenna or several antennas, and is first caught in the far-field region [21]. To ensure maximum power transfer, it needs an impedance matching circuit. The rectifier then transforms from DC to RF power. The output voltage of the rectifier to power a wireless device is frequently too low. To boost

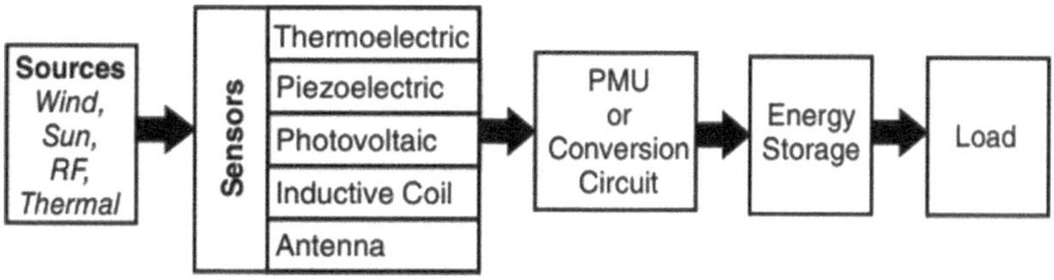

Figure 6.3 Block diagram of an energy harvester.
Reprinted with permission from S.K. Divakaran, et al., RF energy harvesting systems: an overview and design issues, International Journal of RF and Microwave Computer-Aided Engineering 16 (October) (2018).

the DC voltage level, a voltage multiplier is utilized. Finally the harvested energy, in an energy storage device, can be stored in the form of a supercapacitor or rechargeable battery, or it can be connected to the load directly [22]. In contrast to magnetic resonance coupling and inductive of the antenna in the near-field region, RF harvests signal energy in the far-field region of the antenna, an RF energy-harvesting device. In this chapter, RF energy harvesting system evaluates the performance, using the overall efficiency metric which may be done. An RF energy-harvesting system of the overall efficiency ($_{\text{overall}}$) is characterized as follows [23]:

$$_{\text{overall}} = \text{DC output power/RF input power}$$

The DC ratio between the given RF input power and the load entering the system is the overall efficiency. Many factors influence this, including operation voltage, frequency band and load impedance, and transmission power. In terms of the input power, this allows the harvested power to be represented [24].

The RF energy-harvesting approach has some important characteristics that set it apart from other methods. The major characteristics that distinguish RF energy harvesting from other energy sources are reliability and affordability. Reliability is defined as gathering enough energy to allow sensor nodes in any sensor network to operate continuously at any time and in any location. Because RF energy sources including TV networks, radio networks, Bluetooth signals, cellular networks, and Wi-Fi are available both indoors and out, ample and continuous energy may be captured for RF-powered wireless sensors [25]. Wireless sensors fueled by other energy sources are more expensive than those driven by RF energy. The fundamental reason for this is that other energy-harvesting approaches necessitate power conversion modules that are not fully integrated and compatible with normal CMOS processes to harvest their respective energies. The total cost of the energy-collecting system rises because of this flaw. RF energy-harvesting systems, on the other hand, are fully interoperable and integrated with other wireless sensor systems on a single chip. Radio transmissions in the form of electromagnetic radiations carrying energy with frequencies ranging from 3 kHz to 300 GHz are used in RF energy harvesting. The power is transferred using one of two WPT techniques: the near-field (nonradiative) approach [26–28] and (ii) far-field (radiative) approach. The generalized form of the near-field

WPT approach is depicted in Fig. 6.4A, which is further separated into two categories: (a) magnetic resonance coupling and (b) inductive coupling. The magnetic field transfers energy between the receiver and transmitter coils in inductive coupling. The energy is exchanged between the two resonators generated by attaching a capacitor to an induction coil in magnetic resonance coupling. Power strength is proportional to the distance between the coils in the near-field WPT approach, and power transfer declines dramatically as distance increases (signal is attenuated by 60 dB per decade of the distance in free space). To transfer the greatest power between the coils, the transmitter and receiver coils must be aligned also. As a result, near-field approaches will struggle to scale in the IoT architecture, where wireless sensor nodes acquire power from a variety of outdoor and indoor sources. Far-field WPT, on the other hand, uses electromagnetic radiation beams like lasers and microwaves to transfer power over a longer distance (Fig. 6.4B). The signal's power intensity is diminished by 20 dB every decade of distance in free space in the far field. The main goals of energy harvesting technologies are as follows [29]:

- Maintenance or/and performance improvements;
- Elimination or decrease of reliance on batteries;

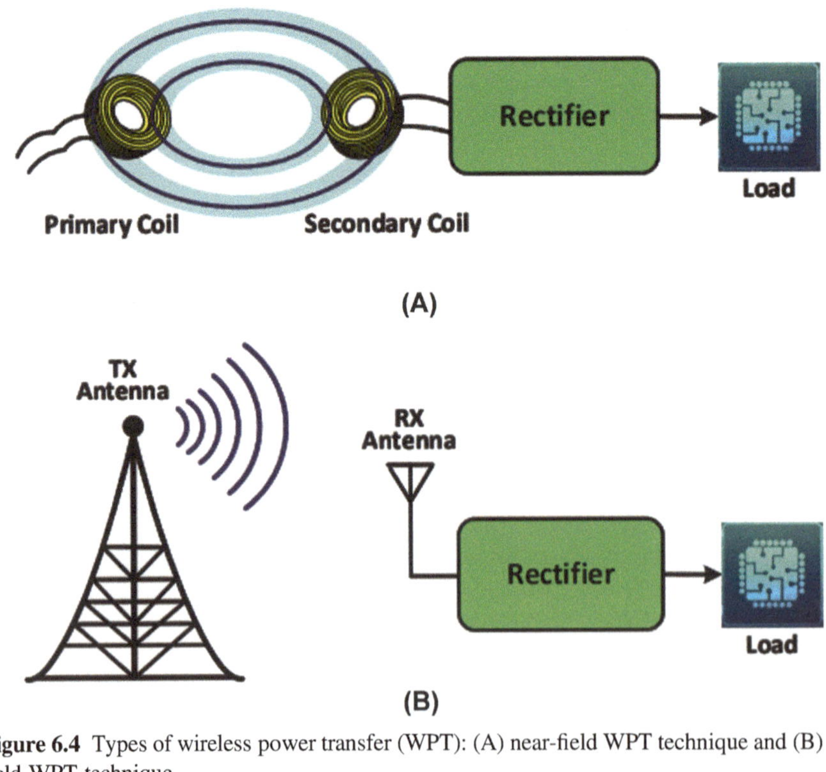

Figure 6.4 Types of wireless power transfer (WPT): (A) near-field WPT technique and (B) far-field WPT technique.
Reprinted with permission from D. Khan, et al., A survey on RF energy harvesting system with high-efficiency RF-DC converters, Journal of Semiconductor Engineering 1 (1) (2020).

- ❖ Ease of installation;
- ❖ Increased lifespan;
- ❖ Removal of main supply lines;
- ❖ Cost savings;
- ❖ Waste reduction.

2.1 RF energy sources in India

AM transmission: 540–1600 kHz (transmitted power few hundred kW); FM radio systems: 88–108 MHz; mobile cellular handsets (transmitted power 1–2 W); Wi-Fi: 2.45 and 5.8 GHz; cell tower transmission of 10–20 W perc; and TV broadcast: 180–220 MHz (transmitted power of few tens of kW) are the most common RF sources in India that enable commercial technology and frequency bands used in cell tower broadcasts in India. As of June 30, 2013 [30], India, with 864.72 million wireless connections, had 747,917 base transceiver stations, and was the country with the second-largest number of mobile phone subscribers [31,32]. This cell tower broadcasts 365 days a year, 24 h a day, and can be utilized as a constant supply of electricity.

3. Matching circuit and antenna

The effects of the matching circuit and the antenna of an RF energy-harvesting system and the performance are discussed in this section. Separately, we discuss impedance-matching circuits, antenna-matching circuit efficiency, and antennas. An antenna is a critical component in RF energy-harvesting systems. Antennas are characterized by antenna gain, frequency range, polarization, physical dimensions, radiation pattern, and application region, among other factors. Some antennas, for example, may use particular frequency bands like UHF and VHF. Antenna types include aperture, array, loop, dipole, horn, log-periodic antennas, and microstrip [33]. The antenna gain is a measurement of how well in a certain direction an RF signal is converted into an electrical signal. High-gain antennas are preferred in terms of an antenna which gains the amount of captured energy which in turn boosts the conversion efficiency. The radiation from an antenna can be isotropic or directed. When the location of an RF signal source is known to improve the energy quantity captured, a directional antenna can be used, or an isotropic antenna can be utilized. The polarization determines the direction of an electric field of an antenna at certain measuring places. When they have the same polarization, the conversion efficiency improves in both the receiving and transmitting antennas. The antenna polarization types are vertical, horizontal, elliptical, and circular [34]. The physical dimensions of the antennas can change depending on the circuit's power needs for its application. Very large antennas are required in radio telescopes to observe astronomical objects, but small antennas, for example, are required in biomedical applications. Spacecraft, ships, planes, and automobiles all use various antenna types. Tools for designing other circuit components or an antenna include Advanced

Design System (ADS), High-Frequency Structure Simulator (HFSS), and Computer Simulation Technology (CST). These programs may compute and optimize antenna parameters such as gain, return loss, and directivity, then build the proper antennas on a printed circuit board.

Multiple antennas can capture more power, which can improve the RF-DC conversion efficiency. However, utilizing multiple antennas increases the circuit size and costs. A triple-band modified Yagi-Uda patch antenna was designed, simulated, and constructed for the 2.45 GHz, 1800 MHz, and 900 MHz frequency bands. The reflected coefficients measured and modeled values were in good agreement. Antenna gains were 3.26, 3.02, and 6.88 dBi for the 2.45 GHz, 1800 MHz, and 900 MHz frequency bands, respectively [35]. Those multiband RF energy harvesters had 15% higher conversion efficiency than single-band RF energy harvesters that were discovered from this study. Using polarization matching and high-gain antennas, the quantity of energy gathered can be enhanced even more. It's worth noting that increasing the efficiency improves the amount of gathered energy, but this does not always imply an improvement in the efficiency. For example, in an energy-harvesting system, the gathered energy amount can be increased with the help of utilizing the same frequency band of two of the same antennas operating, but the system's efficiency remains the same.

For decades, antennas for communication have been explored. In the early stages of development, power-harvesting antennas were still in use [28]. Antenna classification used to be based on design features and uses. Printed planar antennas, aperture antennas, reflector antennas, and wire antennas were originally included. Fig. 6.5 depicts a few different types of antennas. To date, technological advancements have allowed for a range of fabrication procedures and antenna designs to be developed to make them more mature and compact.

For small and compact applications, on-chip antennas are chosen. Plate antennas are common and have various uses. Many recent publications have focused on multiband and wide-band antennas. Narrow-band antennas have been shown to have great energy conversion efficiency; however, they can only collect a limited quantity of energy. Wide-band or multiband fz antennas, on the other hand, can extract more RF energy in space. Low overall efficiency and a big aperture are, however, the downsides. A resonance frequency of antennas having 5.9 and 4.9 GHz, respectively, was constructed with 64.8% and 65.2% PCEs. Lu et al. conducted further research [38] on polarization antennas that backs up the claim that increasing an antenna's bandwidth increases the quantity of power gathered. With three distinct modes, the broadband polarization antenna demonstration in this work enables the antenna to function across a large frequency range. By switching the diodes on and off, and so adjusting their resonance frequencies, the management of the antenna configuration is a common technique in the aforementioned works. This antenna, however, is unable to resonate at two frequencies at the same time because it utilizes separate modes for each frequency. The antenna shown, on the other hand, can simultaneously operate at 5.8 and 2.45 GHz, delivering a PCE of 65% with 2.6 V output and 10 mW/cm^2 power density [39].

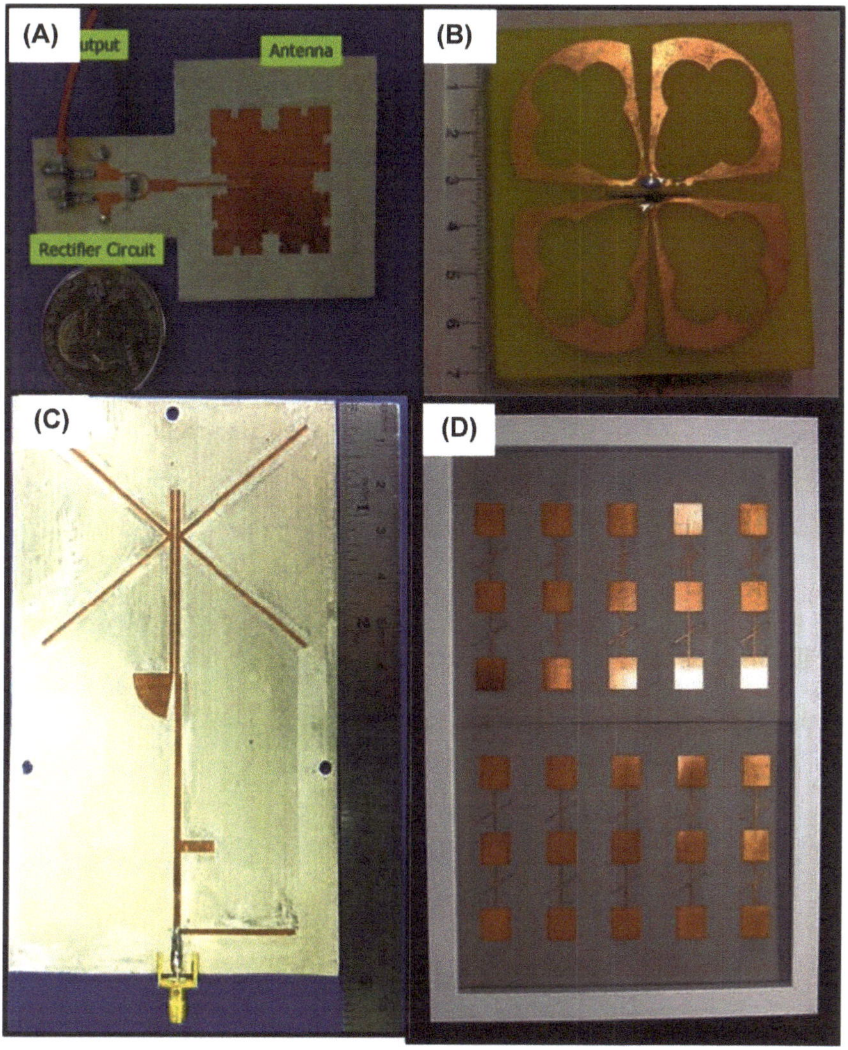

Figure 6.5 (A) 2.45 GHz patch antenna with rectifier, (B) planar dual-polarized antenna, (C) microstrip antenna, and (D) array of stacked differential patch antenna.
Reprinted with permission from U. Olgun, C.C. Chen, J.L. Volakis, Wireless power harvesting with planar rectennas for 2.45 GHz RFIDs. In: 2010 URSI International Symposium on Electromagnetic Theory (EMTs), 2010, pp. 329–331. C. Song, Y. Huang, J. Zhou, J. Zhang, S. Yuan, P. Carter, A high-efficiency broadband rectenna for ambient wireless energy harvesting, IEEE Transactions on Antennas and Propagation 63 (2015) 3486–3495; T. Matsunaga, E. Nishiyama, I. Toyoda, 5.8-GHz stacked differential rectenna suitable for large-scale rectenna arrays with DC connection, IEEE Transactions on Antennas and Propagation 63 (2015) 5944–5949.

4. Photovoltaic energy harvesting

In PV energy scavenging, electricity is generated by equipment employing any other artificial source of light or sunlight. The PV cells are made up of semiconductor materials; from an external source, it absorbs light in most cases. From the semiconductor, absorbed light releases electrons due to the p−n junction's action. The electrodes gather the liberated electrons and holes, resulting in a voltage difference.

The main downside of this technology is that without a light source it cannot function, which limits the number of applications [40]. For commercial purposes, industrial application of PV technology has been successfully utilized in large-scale renewable energy and small-scale energy harvesting. It is utilized in small-scale industries as a portable power bank and leisure trickle charger. On the other hand, in large-scale industry, it is utilized in solar farms and supplemental power generation. Scientists have concentrated their efforts on producing organic thin-film solar cells in recent years which need less energy to manufacture and are less expensive.

For energy harvesting, PV cells that are specially designed can be utilized in both outdoor and indoor settings. In interior spaces, the light intensity of the outdoors is typically lower. The light intensity emitted by the Sun is very much greater than that of a fluorescent tube, incandescent light bulb, or halogen lamp. As a result, the spectral qualities of solar light must be examined to produce the highest practical power; these spectral features govern the operational range of the light type [41]. Because of this, the PV cell efficient performance, on the material which is dependent on where it is constructed, in a specific wavelength range is larger. For a PV harvesting system, a general model proposed is shown in Fig. 6.6.

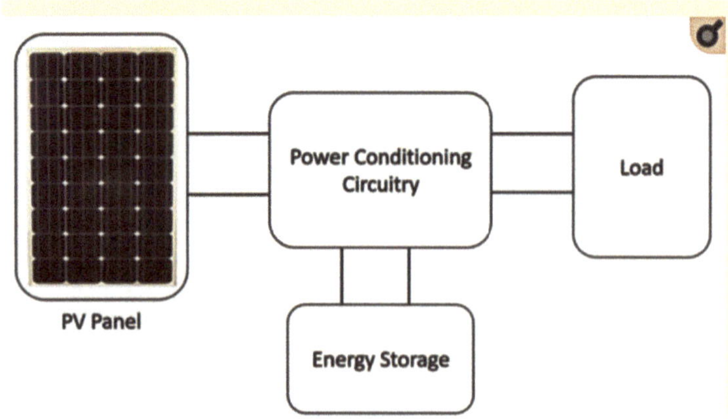

Figure 6.6 Photovoltaic energy-harvesting system.
Reprinted with permission from S.M. Demir, F. Al-Tudjman, A. Muhtaroğlu, Energy scavenging methods for WBAN applications: a review. IEEE Sensors Journal 18 (2018) 6477−6488.

4.1 Hybrid energy harvester

Energy from two separate sources, such as the Sun and radiofrequency energy, can be collected by energy-harvesting systems like hybrid energy harvesters. It has designed to produce more energy than a single energy harvester. A solar/RF energy harvester was developed by Kyotori which produces more electricity than a narrow-band RFEH device [43]. A solar RF hybrid energy harvester is shown in Fig. 6.7.

The antenna performance is influenced by the solar cell's dc connections. When compared to other harvesting circuits, at the output decays the quantity of power received due to solar radiation variations, and the enormous antenna area. With the F antenna, the solar cells are also incorporated, and they serve as the antenna's ground plane [45]. It does, however, exhibit performance in a narrow band.

4.2 Thermoelectric energy harvesting

For climates with temperature variations, thermoelectric energy harvesters are more suited. The temperature difference between the material terminals can be used to efficiently convert energy, while the heat flow provides power. The Carnot theorem and the poor efficiencies of thermoelectric materials mean that even when the heat flow is large, these devices do not supply enough energy. As a result, the best thermoelectric materials are strongly doped semiconductors. The three mechanisms by which thermoelectric devices work are the Thomson effect, Peltier effect, and Seebeck effect [46]. The Seebeck effect is one of the most important thermoelectric power production phenomena, explaining how a circuit made up of two distinct materials can generate a

Figure 6.7 Solar/RF hybrid energy harvester.
Reprinted with permission from K. Niotaki, A. Collado, A. Georgiadis, S. Kim, M.M. Tentzeris, Solar/electromagnetic energy harvesting and wireless power transmission, Proceedings of the IEEE (2014).

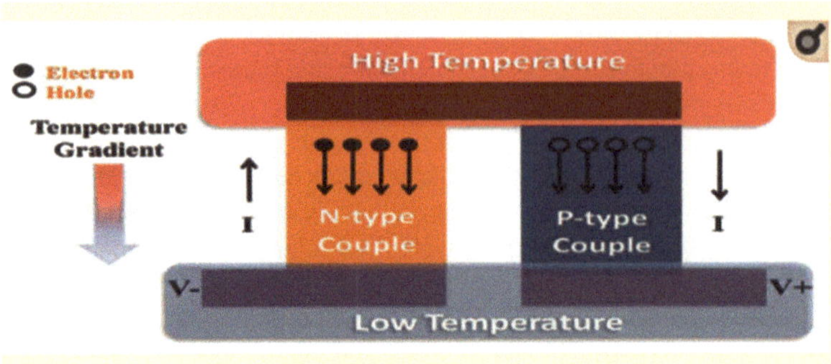

Figure 6.8 The basic operation of a thermoelectric generator (TEG).
Reprinted with permission from S.M. Demir, F. Al-Tudjman, A. Muhtaroğlu, Energy scavenging methods for WBAN applications: a review. IEEE Sensors Journal 18 (2018) 6477–6488.

potential difference or an electromotive force. For the maintenance of the connections between the two materials, different temperatures are used. The advantages of thermoelectric devices are numerous: they are solid-state, they do not emit any electric sound or flashing, they do not produce dust or other particles, they can be oriented in any direction, and they have no moving parts, and so they are highly reliable, acoustically silent, and electrically safe. Furthermore, they are light and compact [47]. Moreover, the biggest disadvantage of this much-hyped technology is that it has a lower rate of conversion, which ranges between 5% and 10% of 60 $\mu W/cm^2$ standard power density. When additional energy-harvesting techniques are included, however, such a disadvantage has little impact on the application. The thermoelectric generator's (TEG) basic function is depicted in Fig. 6.8.

5. Load or energy storage

To generate a specific amount of energy, energy-harvesting circuits are made. To supply the load continuously, an energy-harvesting circuit is needed and the power consumption of the load (e.g., device or sensor) is less than the average captured power. As a result, there is no need for a storage device such as for energy or load. The captured energy in a device should be stored that can store it [21]. The energy-harvesting circuit's impedance is determined by several factors, including the voltage multiplier topologies or rectifier, load impedance, operating frequency, input power level, and so on. Appropriately when the load impedance to the circuit matched can maximum power transfer be achieved? As a result, each operating frequency and input power level has optimal load impedance. Fig. 6.9 shows that the PCE of the highest RF-DC was only reached in particular load resistances, which were variable in the input power levels.

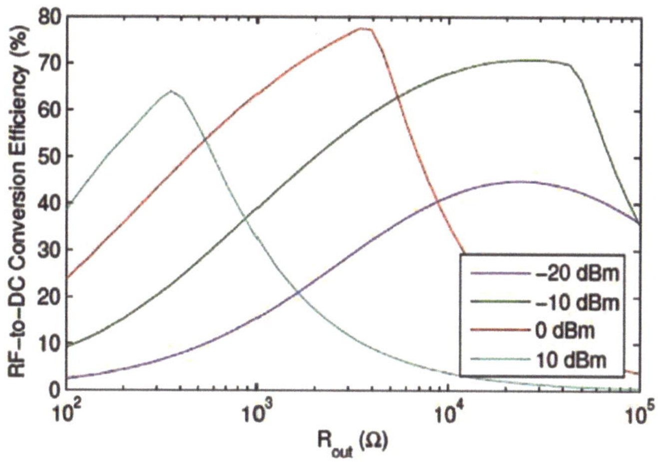

Figure 6.9 Conversion efficiency versus load resistance of RF-DC power.
Reprinted with permission from R. Marshall Blake, M. Morys Marcin, D. Durgin Gregory, Parametric analysis and design guidelines of RF-to-DC Dickson charge pumps for RFID energy harvesting, in: IEEE International Conference on RFID (2015), p. 32e9.

The load resistance value is too low or too high if the RF-DC PCE suffers substantially. The energy-harvesting circuit that should be constructed has to be selected with the optimal load to get the best PCE. Energy storage devices for energy-harvesting systems could be considered. A Ragone plot, which plots energy density versus power density, is used to compare the performance of energy storage systems. The energy density is shown in the Ragone plot, whereas the power density shows the amount of power per mass (W/kg) and reflects the amount of energy per mass (Wh/kg). As a result, the Ragone plot compares the time required to supply energy in energy storage technologies. Rechargeable batteries or supercapacitors may be utilized in RF energy-harvesting devices to store captured energy.

6. RF-DC converters

The methodologies utilized for RF-DC converter hardware circuit designs are discussed in this section. An RF-DC converter is a key block in an RF energy-harvesting system, which converts the RF power acquired by the receiving antenna into useable DC power. For circuit implementation of RF-DC converters, a variety of methodologies have been used, including technology-based techniques and CMOS-based techniques [49]. The RF energy harvester circuit design utilizes HSMS and Schottky diodes, which are traditionally technology-based methodologies because of their low threshold voltage. Furthermore, recent improvements to this technology, such as tunnel diodes, spin diodes, and MIM diodes, have made it more mature. Tunnel diodes, for example, can function at very high frequencies because of their low parasitic components. MIM diodes may also now be combined with the CMOS process, which was previously unachievable with Schottky diodes.

Furthermore, when compared to HSMS and Schottky diodes, spin diodes have a lower threshold voltage. Certain parameters must be taken into account while selecting a diode for rectifier design. Schottky diodes, for example, have high zero-bias junction resistance at microwatt input power levels, which reduces the rectifier's PCE [50]. If the input power level is less than 1 μW, however, due to their low parasitic losses, W-Band ZBD and SMS-7630 diodes are employed, thanks to recent advancements in this technology.

Because of its customized aid in designing and supplying greater sensitivity at lower input voltage than Schottky diodes, CMOS technology has recently received a lot of attention and has been widely published about. The active-circuit technique and the passive-circuit method are two methodologies used in CMOS technology. The active-circuit approach necessitates the use of an external battery for the circuit to operate, and is most commonly employed in active RFIDs or active sensors. The rectifier described uses an external battery to provide a bias voltage using an active technique. The bias voltage is applied to each transistor (which acts as a rectifier) to lower the threshold voltage and improve the rectifier's performance. The external battery, on the other hand, increases the expense and maintenance of the active approach, making it unsuitable for long-term harvesting applications. Passive-circuit techniques, on the other hand, do not require an external source for rectifying the device threshold voltage, but they do necessitate additional circuits. Using an auxiliary block, the rectifier reported creates a threshold compensating voltage. The auxiliary block, usually made up of capacitors and transistors, generates compensating voltage for the main rectification chain's rectifying devices [51]. The rectifying device's threshold voltage has a significant impact on the rectifier's performance. The threshold voltage is defined as the minimum voltage required to switch on the transistor and enable the rectifier to operate. As a result, in the rectifier, the number of stages must be carefully determined because it has a direct impact on the rectifier's performance. Although the output voltage of a rectifier can be increased, the number of stages by increasing the PCE of a multistage rectifier is reduced due to the greater voltage drop across the transistors. As a result, in the PCE and rectifier, the number of stages is mutually exclusive. The PCE of the rectifier decreases as the number of stages of the rectifier increases, and vice versa. To improve the rectifier's performance, some methods have been suggested to lower the transistor's threshold voltage.

The fundamental problem with scavenging RF power from ambient sources is the rectifier's restricted signal intensity at the input. Unpredictable attenuations in signal strength over distance occur because of path loss, the power source, regulatory bodies, antenna orientation, maximum signal strength limits, and the transmission medium in which the rectifier is used and the presence of hurdles/obstacles between the rectifier; therefore, the signal strength is limited. As a result [52,53], in an impedance matching circuit, although it has been boosted, RF voltage levels are not sufficient for the rectifying components in the rectifier, the overdrive voltages are generated to have low conduction losses. As a result, the rectifier is unable to capture the maximum amount of energy feasible, and its performance suffers. The main problem is to develop a rectifier that can create, from minimal input RF power, a battery-like voltage. A rectifier can be configured in three different ways: (a) diode-based, (b) voltage multiplier, and

(c) bridge of diodes. A rectifier circuit's most important component is the diode. The diode's conduction resistance, junction capacitance, and saturation current are all important factors in the rectifier's performance [54]. The PCE of the rectifier circuit is determined by the diode's performance. Rectenna diodes made of silicon Schottky are the most common.

7. Energy storage device efficiency

The impedance of the load block or energy storage device is proportional to its efficiency. The load block's impedance or energy storage device's matching efficiency can be calculated. There is no reflection coefficient and reflected power which is zero since no impedance mismatch between the load and the voltage multiplier circuit [55]. As a result, the circuit has no power loss.

7.1 Batteries

In energy-harvesting circuits, batteries can also be used as a rechargeable power supply and for energy storage as an alternative to supercapacitors. Nickel metal hydride (NiMH), lithium-ion, nickel-cadmium (NiCd), lithium (Li), and sealed lead-acid (SLA) are examples of different types of batteries [56]. The lithium and lithium alloy batteries are more advantageous and are more efficient than conventional batteries. With a series connection, internal resistance and an ideal voltage source can be presented as the equivalent model of a battery. In an energy-harvesting circuit, when a battery is used as a load, the charging current increases and its charging efficiency changes. To enhance the charging efficiency of a battery, there is an ideal charging current. Batteries have a higher energy density than regular capacitors and supercapacitors as a benefit. On the other hand, supercapacitor batteries have a shorter lifetime and lower power density. Furthermore, a storage device known as a supercapatteries is a new energy device that has recently been invented. Redox materials and carbon nanotubes are used as electrodes in supercapatteries, the greatest features which combine batteries and supercapacitors. It has a larger energy storage capacity than a supercapacitor and can discharge and charge more quickly than a battery [57]. As a result of these benefits, it will contribute to the efficiency of energy-harvesting systems. The type of energy storage device used in an energy-harvesting system should be chosen based on its benefits and drawbacks, such as leakage current, longevity, size, power density, discharging efficiency, charging, and energy density.

7.2 Supercapacitors

Supercapacitors are high-energy storage electrochemical double-layer capacitors. They have a higher power density than batteries and higher energy density than traditional capacitors. Supercapacitors are used for functions between those of ordinary capacitors and batteries because of these characteristics. Supercapacitors are preferred for storing gathered energy due to their long lifetime good discharge performance, and short charging times. A perfect capacitor and an internal resistance can be used to model a supercapacitor. In 10 years, the internal resistance doubles, and the capacity of a supercapacitor reduces to 80%. The supercapacitor's actual life might decrease in 20 years as their capacities decrease. The charging and discharging efficiencies of supercapacitors determine their efficiency. When a supercapacitor is used as a load, the charging efficiency, energy-harvesting circuit, drops as the internal resistance increases [58,59]. The supercapacitor has a high charging efficiency because of its low internal resistance. In mU levels, a supercapacitor with a 350 F internal resistance capacity is measured. However, the high self-discharge feature reduces the supercapacitor efficiency, lowering the energy-harvesting system's efficiency.

7.3 Power management

The common process that includes energy storage units, a voltage multiplier, and rectifier is known as power management. It has an impact on the efficiency of all of these units in this setting. Power management is a smart application in a wireless energy-harvesting system, which maximizes captured energy and efficiency [60]. For these reasons, many energy-harvesting circuits include a power management unit (PMU).

The PMU is shown in Fig. 6.10 in an energy-collecting circuit. Maximum power point tracking, with time-varying power which is applied to energy sources, can

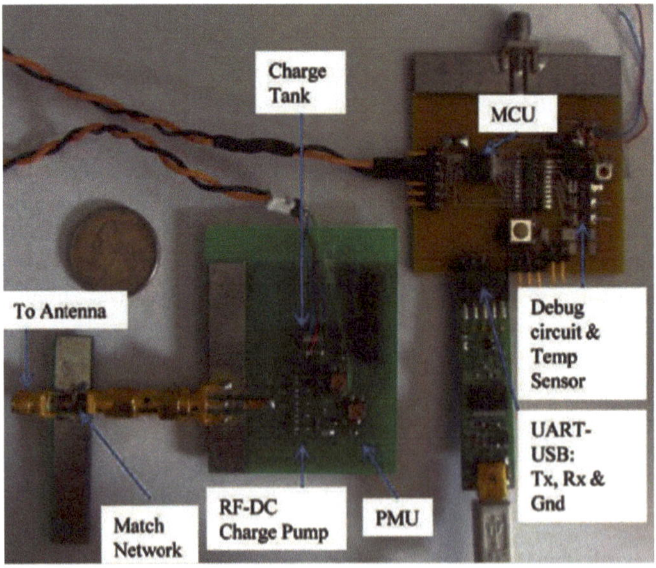

Figure 6.10 Energy-harvesting circuit—PMU.
Reprinted with permission from J. Vyas Rushi, B. Cook Benjamin, Y. Kawahara, M. Manos, Tentzeris, EWEHP. A batteryless embedded sensor-platform wirelessly powered from ambient digital-TV signals, IEEE Transactions on Microwave Theory and Techniques 61 (6) (2013) 2491e505.

provide power control. The MPPT seeks to obtain the most power by altering the current or voltage based on the current–voltage curve of the energy source. The MPPT algorithm, which operates under the PMU application, for maximum generated power is used to discover the ideal position. Furthermore, the MPPT algorithm can be implemented using a variety of methods. In the energy-harvesting circuit, the PMU enables the wireless to operate and protect energy storage devices such as capacitors and batteries and to monitor collected energy levels. The energy-harvesting system's efficiency may be tracked and optimized with the help of the PMU. The efficiency of the DC–DC power converter was improved with the use of an optimized PMU [62]. For the efficiency of the DC–DC power converter, fitness function, on-time, and the inductor value were all optimized. The optimum parameter values were determined using particle swarm and resistor emulation optimization approaches. At low power levels, to follow the maximum power point, the resistor emulation technique was utilized. The goal of the particle swarm optimization methodology, on the other hand, was to identify the optimal inductor and on-time values for optimum efficiency. The improved PMU was shown to be 9.25% more efficient than the traditional PMU in this investigation. At RF energy harvesting, a power management circuit is used, with an efficiency of 87.7% for a power of 2.4 mW. The harvested energy amount is greatly increased for RF energy harvesting by using an adaptive optimization algorithm and MPPT in a microcontroller-based power management system [63]. As a result of wireless energy-harvesting systems, power management can be made more efficient. The PMU's efficiency can be used to calculate the voltage multiplier efficiency, energy storage units, and rectifier.

8. Recent trends in rectennas

Due to the widespread deployment of Wi-Fi routers, the majority of published studies focus on 5.8 and 2.45 GHz on ISM bands, and 2100 MHz, GSM 900, and GSM 1800 frequencies are all covered [64]. Rectennas operating at 94 and 24 GHz, which enables millimeter-wave communications, have also been reported as a result of the expansion toward 5G in recent developments and wireless communication. Compact, omnidirectional, and high-gain antennas, such as frequency-independent antennas, substrate-integrated waveguide antennas, dielectric resonator antennas, and microstrip antennas, are commonly used for rectennas. Those antennas further using artificial materials are modified to reach high efficiency, small form factor, and wide bandwidth. Fig. 6.11 shows the most linearly polarized antennas to reduce mismatch losses which are converted to circular polarization.

Rectennas are made using low-loss substrates, such as Rogers R4003C, RT Duroid 5880, glass, and Arlon, to achieve higher efficiency than FR4. It has a bow-tie cross-dipole planar antenna with dual linear polarization, 400 MHz broad bandwidth, bidirectional radiation pattern, and omnidirectional radiation pattern. The bow-tie antenna is further developed to provide a considerably broader bandwidth and linearly polarized cross dipole antenna into a frequency-independent antenna. The proposed

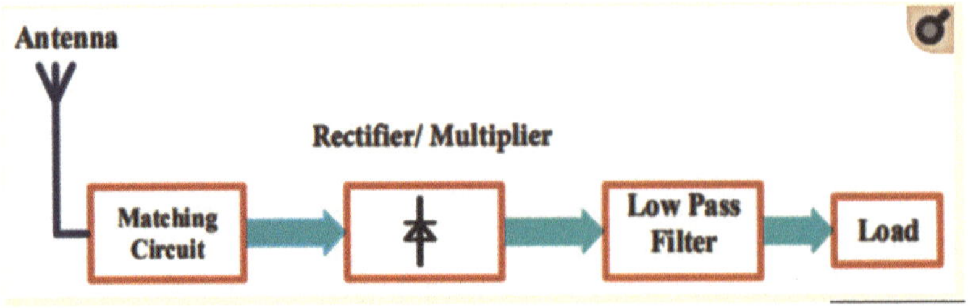

Figure 6.11 A radio-frequency energy-harvesting (RF-EH) circuit block diagram. Reprinted with permission from S.K. Divakaran, et al., RF energy harvesting systems: an overview and design issues, International Journal of RF and Microwave Computer-Aided Engineering 16 (October) (2018).

rectenna is only designed for a 5 dBm input power and does not account for load changes. It operates at frequencies of 2.45, 2.15, 1.85, 0.9, 0.7, and 0.55 GHz, with 67% efficiency at 5 dBm input power [66]. L-Probe microstrip patch has two single-port antennas with neighboring ground fabrication that are stacked back to back. To boost bandwidth, over the ground the patch is suspended with a thick air gap between. To combine the rectifier with the antenna, an L-probe feed is employed. At frequencies of 0.8–2.3 GHz, this structure resonates. However, only 0.9, 1.8, and 2.4 GHz are resonant in its rectifier circuit. At 10 dBm input power and a 0.9 GHz working frequency, it has a maximum efficiency of 45%. Because the entire output voltage can be raised by merging many ports, the multiport rectenna idea is the most recent trend [67,68]. The Shen 90s three-port pixel rectenna at an input power of 19 dBm band achieved a 19% efficiency working in the GSM 1800 MHz. On Rogers RO4003C substrate, it is made up of a rectangular grid that is manufactured with 15 × 15 square pixels. DC combining from three ports use is allowed for wide beamwidth, allowing power to be received from all directions, despite the antenna's huge overall size and lower efficiency at low-input levels because of mutual coupling losses. At −15 dBm input power, at 2.1, 2.4, and 3.3 GHz of differential the operating rectenna had 53%, 31%, and 15.56% efficiency, respectively. Multiband features were attained that are electromagnetically using a two-layer square antenna connected through slots [69]. At the bottom layer, it is supplied differently. To obtain high gain, a large reflector is utilized on the antenna's backside, of increased overall size although this comes at a cost. For rectenna feeding, CPW transmission lines are a good choice because of their low dispersion, low radiation loss, and ease of passive and active device integration. It also aids in the attainment of a wide bandwidth for antennas and allows for rectifier impedance easy matching. The most popular choice for rectennas is patch antennas since they are lighter, less expensive, and easier to integrate than dipole and monopole antennas. To construct an impedance variation independent of the frequency with broad bandwidth antennas, frequency-independent antennas or spirals are recommended over YagiUda [70,71]. To accomplish compactness and circular polarization, slot antennas are employed. Because of their capacity to concentrate, metamaterial-based antennas have high efficiency and radiation, and reduced surface waves.

8.1 Methods to design an efficient RFEH system

Individual component efficiency determines the total PCE of an RFEH system. By optimizing individual efficiencies and merging them with the least amount of loss, maximum efficiency can be achieved [72]:

- The system's efficiency is determined by the appropriate frequency and range of operation. The rectifier topology to be used is determined by the needed output power.
- Use a boost converter between the antenna and the rectifier circuit if the input RF voltage is low. If you want to get a combined DC output from several antennas or rectennas, utilize a DC combining circuit [73].
- A small matching network that can match the antenna impedance to the rest of the circuit regardless of input power, frequency, or load fluctuation.
- The needed output voltage depending on the application, many stages, and the topology of the rectifier must be chosen.
- A compact circularly polarized antenna with wide bandwidth and beam width for receiving maximum RF power with the least amount of loss should be used.
- To enhance the output, DC voltage uses a voltage booster if it is too low.

9. Applications

Energy-harvesting technology is making inroads into some hot sectors, including Big Data, IoT, and 5G communications, offering practical solutions for long-term, large-scale, wireless data exchange systems and low-cost environmental monitoring [74]. Potential uses of energy harvesting and several existing wireless communications are presented in this section.

9.1 Healthcare of animals

Animals pretend to be fine even when they are sick and hide their illness, according to studies. When evident sickness symptoms show, for doctors, this trait makes it more difficult to save the animal's life. As a result, how to keep track of the health of a large number of animals in a farm or zoo becomes a vital but difficult task. The multitier RF-EHWC network is a cost-effective option, as demonstrated in Fig. 6.12.

Creating an ad hoc BAN Tier 1 involves implanting a BAN into an animal's body as energy-harvesting biosensors. In Tier 2, to build a local area healthcare network, each BAN is linked to a sink node. In an animal's leg band, ear tag, or collar, the sink node is embedded. The data collected are sent from the biosensors to the nearest access point [76]. Tier 3 establishes a wide-area healthcare network to the Ethernet for real-time health monitoring by connecting all access points.

9.2 Wearable devices

Wearable technologies, such as a smart bracelet or smartwatch, have gained significant interest in recent years for tracking physical activity and health monitoring. As a

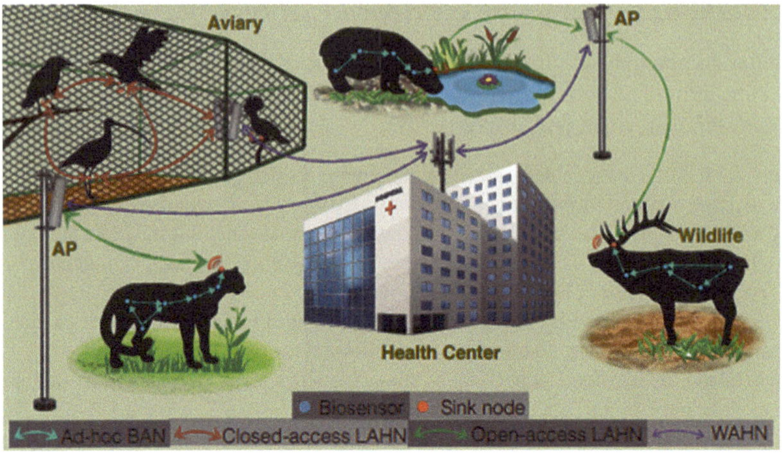

Figure 6.12 A three-tier architecture of an RF-EHWC network for the healthcare of animals. Reprinted with permission from Y. Luo, L. Pu, Y. Zhao, RF Energy Harvesting Sensor Networks for the Healthcare of Animals: Opportunities and Challenges (2018).

self-contained ecosystem, commercial wearable devices are frequently designed that combine data gathering, message display into a single device, wireless connection, and processing. Today's wearable devices rely significantly on high-capacity batteries because of their high energy consumption. Suppose we detach the information display modules and data processing from the wearable devices and merely combine data transmission and sensing, the energy consumption of the wearable devices would be considerably decreased, with renewable energy sources allowing them to be charged. A researcher used a specialized energy source in an experiment, to charge wearable equipment held in the right hand. The energy source's transmission power and frequency bands were 464–468 MHz and 2 W, respectively. Based on the trial results, the device attached to the rear of the head received the least power, averaging 12 dBm; the wearable device attached to the front of the head was able to harvest the most power, averaging 6.9 dBm [77,78]. A hybrid energy harvester can use a wearable device to collect extra energy for reliable operation. Thermal and radiofrequency energy are recommended as energy sources. A tiny TEG is a device from the human body that gathers heat energy and produces less than 100 mV output voltages. An ultralow-voltage multiplier is required from 100 mV to 1.35 V to enhance the low output voltage. The applied voltage multiplier, on the other hand, requires a 600-mV voltage to start up. To capture radio energy and produce the requisite start-up voltage, an RF rectifier with a sensitivity of 10 dBm is employed.

9.3 5G-assisted RF-EHWC

EM waves due to the substantial spreading loss in free space as well as the power density and safety issues of ambient RF energy are low. The 5G mobile technology could solve this problem to harvest energy more efficiently, with low power density, allowing RF-EHWC networks with greater flexibility. As shown in Fig. 6.13, the antenna size can be considerably decreased by the communication frequency increasing to the mm-Wave band [79,80]. In a 5G network, this permits dense deployment with a huge number of small femtocells of antennae [81,82]. From the following three perspectives, deployment of the antenna array and the high density are advantages of an RF-EHWC system for this implementation.

- ❖ **Low propagation loss**: Ultradense networking is seen as a possible 5G communication technology. The HBS can be deployed in a high-density configuration to reduce the distance between the RF-EHWC devices and the energy source [83,84]. As a result, reduced RF energy propagation loss can be expected in a 5G-assisted RF-EHWC network.
- ❖ **High beam-steering resolution**: The mm-wave phased array is capable of forming a large number of controlled beams with great precision. The difference between the directions of neighboring beams using a 32-element phased array can be as small as 1.4 degrees. The HBS has a lot of flexibility when it comes to allocating energy transmissions and power for information thanks to the beam-steering high resolution. It is even possible to power and communicate data to nearby RF-EHWC devices at the same time. The antenna-switching RF-EHWC technique allows the RF-EHWC device with huge antennas from many HBSs to receive information and collect energy at the same time [85,86].
- ❖ **Large antenna gain**: HBS (femtocells) can form a narrow beam using an array antenna for efficient power transmission or enhance spatial multiplexing gain for high-speed data transfer to increase antenna gain [87,88]. Fig. 6.13 shows both the receiver and the sender equipped with the array antenna; the power of the RF waves arriving at the reception can be 20 dB higher than when only a single antenna is used [89,90].

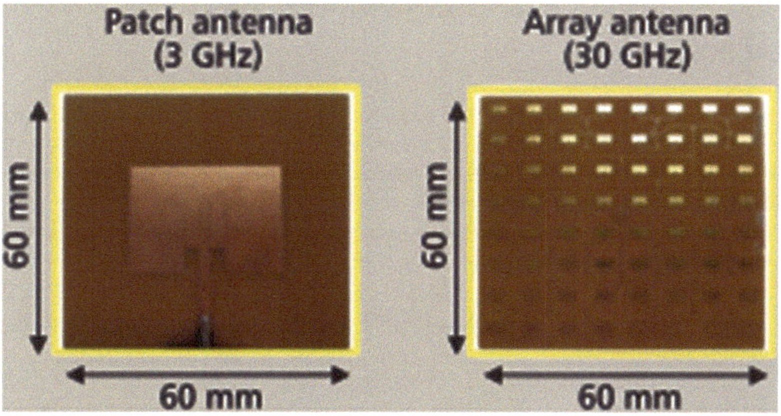

Figure 6.13 Antenna comparisons: Antenna array with 64 (8 × 8) elements for 5G communications in 30 GHz, patch antenna for traditional wireless communications in 3 GHz. Reprinted from W. Roh, J.Y. Seol, J. Park, B. Lee, J. Lee, Y. Kim, J. Cho, K. Cheun, F. Aryanfar, Millimeter-wave beamforming as an enabling technology for 5G cellular communications: theoretical feasibility and prototype results. IEEE Communications Magazine 52 (2014) 106–113.

10. Conclusion

The state-of-the-art in RF power-harvesting technology has been summarised in this chapter. RF power harvesting has already been used in some practical applications. The antenna, IMN, and voltage multiplier are the three major components of a basic RF power-harvesting unit. Harmony in the integration of all components determines the system's overall efficiency. In addition, the chapter has explored the designs and ideas behind each module. RF electromagnetic waves can permeate soft tissues, and are abundant in space and nonharmful. These characteristics make RF electromagnetic waves a viable alternative to batteries in a variety of applications. RF power harvesting, in particular, promotes low-power medical and healthcare equipment, as well as the IoT and WSN development, by allowing for mobility of usage. Furthermore, advancements in the integration of RF power-harvesting circuits into CMOS technology have resulted in the creation of a truly wireless SoC. Aside from recent advancements, in RF power-harvesting technology, there remain several areas that can be improved, such as extending the operation range, reducing system dimensions, optimizing PCE, and reducing transmission loss. Moreover, in severe working environments, research focusing on RF WPH, like underwater zones or implantation circumstances, is gaining a lot of interest to improve the technology's capabilities. To mature this technology, packing procedures and fine manufacturing with cost-effective production are required. In a nutshell, RF WPH technology is becoming a reality slowly but steadily. Even though this technology currently has numerous flaws, overcoming these obstacles could usher in a new era of sustainable and clean energy for the power industry.

References

[1] D. Brunelli, L. Benini, C. Moser, L. Thiele, An efficient solar energy harvester for wireless sensor nodes, in: 2008 Design, Automation, and Test in Europe, 2008, pp. 104–109.
[2] A. Cuadras, M. Gasulla, V. Ferrari, Thermal energy harvesting through pyroelectricity, Sensors and Actuators A: Physical 158 (2010) 132–139.
[3] A. Khaligh, P. Zeng, C. Zheng, Kinetic energy harvesting using piezoelectric and electromagnetic technologies—state of the art, IEEE Transactions on Industrial Electronics 57 (2010) 850–860.
[4] T.H. Lee, The design of CMOS radio-frequency integrated circuits, Communications Engineer 2 (2004) 47.
[5] J. Zhang, Z.P. Wu, C.G. Liu, B.H. Zhang, B. Zhang, A double-sided rectenna design for RF energy harvesting, in: 2015 IEEE International Wireless Symposium (IWS), 2015, pp. 1–4.
[6] U. Olgun, C.C. Chen, J.L. Volakis, Wireless power harvesting with planar rectennas for 2.45 GHz RFIDs, in: 2010 URSI International Symposium on Electromagnetic Theory (EMTs), 2010, pp. 329–331.
[7] C. Hoarau, N. Corrao, J.D. Arnould, P. Ferrari, P. Xavier, Complete design and measurement methodology for a tunable RF impedance-matching network, IEEE Transactions on Microwave Theory and Techniques 56 (2008) 2620–2627.

[8] D. Khan, et al., A survey on RF energy harvesting system with high-efficiency RF-DC converters, Journal of Semiconductor Engineering 1 (1) (2020).
[9] Y.S. Hwang, C.C. Lei, Y.W. Yang, J.J. Chen, C.C. Yu, A 13.56-MHz low-voltage and low-control-loss RF-DC rectifier utilizing a reducing reverse loss technique, IEEE Transactions on Power Electronics 29 (2014) 6544–6554.
[10] H. Sun, G. Xu, A differentially-driven rectifier for enhanced RF power harvesting, in: 2015 IEEE MTT-S International Microwave Workshop Series on Advanced Materials and Processes for RF and THz Applications (IMWS-AMP), 2015, pp. 1–3.
[11] M.P. Praveen, N.B. Mehta, Trade-offs in analog sensing and communication in RF energy harvesting wireless sensor networks, in: 2016 IEEE International Conference on Communications (ICC), 2016, pp. 1–6.
[12] J.H. Lee, W.J. Jung, J.W. Jung, J.E. Jang, J.S. Park, A matched RF charger for wireless RF power harvesting system, Microwave and Optical Technology Letters 57 (2015) 1622–1625.
[13] L.-G. Tran, et al., RF power harvesting: a review on designing methodologies and applications, Micro and Nano Systems Letters 5 (2017) 14.
[14] Y. Li, L. Fu, M. Chen, K. Chi, Y.H. Zhu, RF-based charger placement for duty cycle guarantee in battery-free sensor networks, IEEE Communications Letters 19 (2015) 1802–1805.
[15] M. Piñuela, P.D. Mitcheson, S. Lucyszyn, R.F. Ambient, Energy harvesting in urban and semi-urban environments, IEEE Transactions on Microwave Theory and Techniques 61 (2013) 2715–2726.
[16] D. Bouchouicha, F. Dupont, M. Latrach, L. Ventura, Ambient RF energy harvesting, in: Proceedings of the International Conference on Renewable Energies and Power Quality (ICREPQ), Granada, Spain, March 23, 2010, pp. 1–4.
[17] M. Gorlatova, A. Wallwater, G. Zussman, Networking low-power energy harvesting devices: measurements and algorithms, IEEE Transactions on Mobile Computing 12 (2013) 1853–1865.
[18] N. Michelusi, K. Stamatiou, M. Zorzi, On optimal transmission policies for energy harvesting devices, in: Proceedings of the IEEE Information Theory and Applications Workshop (ITA), San Diego, CA, USA, February 5–10, 2012, pp. 249–254.
[19] N.A. Pantazis, S.A. Nikolidakis, D.D. Vergados, Energy-efficient routing protocols in wireless sensor networks: a survey, IEEE Communications Surveys and Tutorials 15 (2013) 551–591.
[20] S.K. Divakaran, et al., RF energy harvesting systems: an overview and design issues, International Journal of RF and Microwave Computer-Aided Engineering 16 (2018) (October).
[21] S. Kim, R. Vyas, J. Bito, K. Niotaki, A. Collado, A. Georgiadis, M.M. Tentzeris, Ambient RF energy-harvesting technologies for self-sustainable standalone wireless sensor platforms, Proceedings of the IEEE 102 (11) (2014) 1649–1666.
[22] H. Shoki, Issues and initiatives for practical deployment of wireless power transfer technologies in Japan, Proceedings of the IEEE 101 (6) (2013) 1312–1320.
[23] G. Chaudhary, P. Kim, Y. Jeong, J.H. Yoon, Design of high-efficiency RF-DC conversion circuit using novel termination networks for RF energy harvesting system, Microwave and Optical Technology Letters 54 (10) (2012) 2330–2335.
[24] C. Mikeka, H. Arai, Design issues in radio frequency energy harvesting system, in: Sustainable Energy Harvesting Technologies—Past, Present and Future, InTech, Europe, 2011.

[25] H. Hong, X. Cai, X. Shi, X. Zhu, Demonstration of a highly efficient RF energy harvester for Wi-Fi signals, in: 2012 International Conference on Microwave and Millimeter Wave Technology (ICMMT), IEEE, Shenzhen, China, 2012, pp. 1−4.
[26] M. Arrawatia, M. Baghini, G. Kumar, Broadband bent triangular omnidirectional antenna for RF energy harvesting, IEEE Antennas and Wireless Propagation Letters 15 (2016) 36−39.
[27] A. Shameli, A. Safarian, A. Rofougaran, M. Rofougaran, F. De Flavius, Power harvester design for passive UHF RFID tag using a voltage boosting technique, Microwave Theory and Techniques IEEE Transactions on 55 (6) (2007) 1089−1097.
[28] P. Nintanavongsa, U. Muncuk, D.R. Lewis, K.R. Chowdhury, Design optimization and implementation for RF energy harvesting circuits, IEEE Journal on Emerging and Selected Topics in Circuits and Systems 2 (1) (March 2012) 24−33.
[29] G. Papotto, F. Carrara, G. Palmisano, A 90-nm CMOS threshold-compensated RF energy harvester, Solid-State Circuits IEEE Journal of on page(s) 46 (9) (Sept. 2011) 1985−1997.
[30] P. Nintanavongsa, R. Doost, M. Di Felice, K.R. Chowdhury, Device characterization and cross-layer protocol design for RF energy harvesting sensors, Elsevier Journal of Pervasive and Mobile Computing 9 (1) (2013).
[31] S.C. Folea, G. Mois, A low-power wireless sensor for online ambient monitoring, IEEE Sensors Journal 15 (2) (2015) 742−749.
[32] T. Soulos, V. George, Smart antennas for mobile communication systems: benefits and challenges, Electronics & Communication Engineering Journal 11 (2) (1999) 84−94.
[33] N.A. Eltresy, Ph.D. Thesis "Energy harvesting solution for wireless sensors in IoT applications", in: Tech. Rep., Ain Shams University, Faculty of Engineering, Electronics and Communications Engineering, LAP LAMBERT, 2020, 6202796596.
[34] W. Ghonaim, S.A. Taie, A new image segmentation algorithm based on particle swarm optimization and rough set, Journal of Computers 13 (1) (2018) 130−139.
[35] D. Khan, et al., A design of ambient RF energy harvester with sensitivity of −21 dBm and power efficiency of a 39.3% using internal threshold voltage compensation, MDPI Energies 11 (5) (2018).
[36] C. Song, Y. Huang, J. Zhou, J. Zhang, S. Yuan, P. Carter, A high-efficiency broadband rectenna for ambient wireless energy harvesting, IEEE Transactions on Antennas and Propagation 63 (2015) 3486−3495.
[37] T. Matsunaga, E. Nishiyama, I. Toyoda, 5.8-GHz stacked differential rectenna suitable for large-scale rectenna arrays with DC connection, IEEE Transactions on Antennas and Propagation 63 (2015) 5944−5949.
[38] M. Stoopman, S. Keyrouz, H.J. Visser, K. Philips, W.A. Serdijn, A self-calibrating RF energy harvester generating 1 V at 26.3 dBm, in: Proc. of IEEE Symposium on VLSI Circuits (VLSI), 2013, pp. 226−227 (Kyoto).
[39] J.M. Barcak, H.P. Partal, Efficient RF energy harvesting by using multiband microstrip antenna arrays with multistage rectifiers, in: Proceedings of IEEE Subthreshold Microelectronics Conference (SubVT), 2012, pp. 1−3 (Waltham, MA).
[40] Z. Zeng, X. Li, A. Bermak, C.-Y. Tsui, W.-H. Ki, A WLAN 2.4-GHz RF energy harvesting system with reconfigurable rectifier for wireless sensor network, in: Proceedings of IEEE International Symposium on Circuits Systems (ISCAS), May 2016, pp. 2362−2365.
[41] M.T. Penella-Lopez, M. Gasulla-Forner, Radiofrequency Energy Harvesting, in: Powering Autonomous Sensors, Springer, Netherlands, 2011, pp. 125−147.
[42] S.M. Demir, F. Al-Tudjman, A. Muhtaroğlu, Energy scavenging methods for WBAN applications: a review, IEEE Sensors Journal 18 (2018) 6477−6488.

[43] Takhedmit, L. Cirio, S. Bellal, D. Delcroix, O. Picon, Compact and efficient 2.45 GHz circularly polarized shorted ring-slot rectenna, Electronics Letters 48 (5) (2012) 253–254. March.
[44] K. Niotaki, A. Collado, A. Georgiadis, S. Kim, M.M. Tentzeris, Solar/electromagnetic energy harvesting and wireless power transmission, Proceedings of the IEEE 102 (11) (2014).
[45] H. Sun, W. Geyi, A new rectenna with all polarization receiving capability for wireless power transmission, IEEE Antennas and Wireless Propagation Letters 15 (2015) 814–817.
[46] U. Olgun, C.-C. Chen, J.L. Volakis, Investigation of rectenna array configurations for enhanced RF Power harvesting, IEEE Antennas and Wireless Propagation Letters 10 (2011) 262–265.
[47] H. Jabbar, Y.S. Song, T.T. Jeong, RF energy harvesting system and circuits for charging of mobile devices, IEEE Transactions on Consumer Electronics 56 (1) (2010) 247–253. February.
[48] R. Marshall Blake, M. Morys Marcin, D. Durgin Gregory, Parametric analysis and design guidelines of RF-to-DC Dickson charge pumps for RFID energy harvesting, in: IEEE International Conference on RFID, 2015, p. 32e9.
[49] L. Liu, R. Zhang, K.C. Chua, Wireless information transfer with opportunistic energy harvesting, IEEE Transactions on Wireless Communications 12 (1) (2013) 288–300.
[50] H. Nishimoto, Y. Kawahara, T. Asami, A Prototype implementation of the wireless sensor network using TV broadcast RF energy harvesting, in: Proceedings of the 12th ACM International Conference Adjunct Papers on Ubiquitous Computing Adjunct, 2010, pp. 373–374.
[51] V. Kuhn, F. Seguin, C. Lahuec, C. Person, Enhancing RF-to-DC conversion efficiency of wideband RF energy harvesters using multi-tone optimization technique, International Journal of Microwave and Wireless Technologies 8 (2) (2016) 143–153.
[52] D. Bouchouicha, F. Dupont, M. Latrach, L. Ventura, Ambient RF energy harvesting, in: IEEE International Conference on Renewable Energies Power Quality (ICREPQ'10), 2010, pp. 486–495.
[53] A. Sample, J.R. Smith, Experimental results with two wireless power transfer systems, in: IEEE Radio Wireless Symposium, 2009, pp. 16–18.
[54] R.A. Rahim, S.I.S. Hassan, F. Malek, M.N. Junita, M.F. Jamlos, An investigation of ambient radio frequency as a candidate for energy harvesting source, in: 2012 IEEE Symposium on Industrial Electronics and Applications (ISIEA2012), 2012.
[55] N. Abdullah, A.M. Shire, E. Mohd, Rectenna for RF energy harvesting, in: International Conference on Advances in Electrical, Electronic and System Engineering, 2016, pp. 14–16.
[56] P. Kumari, J. Sahay, in: Investigation on RF Energy Harvesting International Conference on Innovations in Power and Advanced Computing Technologies, 2017.
[57] V. Kumar, R. vakeeram, Differential microstrip antenna with sidewall patch for RF energy harvesting, in: 3rd International Conference for Convergence in Technology (12CT), 2018.
[58] M. Pushpavalli, P. Abirami, et al., Implementation of RF energy harvesting for mobile charging, in: IEEE International Conference on Computational Intelligence, 2018.
[59] A. seeks, E. tammam, et al., High efficiency 2.45 GHz low power hybrid junction rectifier for RF energy harvesting, in: IEEE Standard, 2018.
[60] N.A. Eltresy, D.N. Elsheakh, E.A. Abdallah, H.M. Elhennawy, RF energy harvesting using efficiency dual-band rectifier, in: 2018 Asia-Pacific Microwave Conference (APMC), Koyoto, Japan, November 2018.

[61] J. Vyas Rushi, B. Cook Benjamin, Y. Kawahara, M. Manos, Tentzeris, EWEHP. A batteryless embedded sensor-platform wirelessly powered from ambient digital-TV signals, IEEE Transactions on Microwave Theory and Techniques 61 (6) (2013) 2491e505.
[62] X. Lu, P. Wang, D. Niyato, D.I. Kim, Z. Han, Wireless networks with RF energy harvesting: a contemporary survey, IEEE Communications Surveys and Tutorials 17 (2) (2014) 757−789.
[63] J.H. We, S.J. Kim, B.J. Cho, Hybrid composite of screen-printed inorganic thermoelectric film and organic conducting polymer for flexible thermoelectric power generator, Energy 73 (2014) 506−512.
[64] X. Gu, S. Hemour, K. Wu, Enabling far-field ambient energy harvesting through multi-physical sources, in: Asia-Pacific Microwave Conference (APMC), 2018, p. 204.
[65] S. Divakaran, D.K.D. Nasimuddin, RF energy harvesting systems: an overview and design issues, The International Journal of RF and Microwave Computer-Aided Engineering 29 (2018) e21633.
[66] A. Nimo, T. Beckedahl, T. Ostertag, L. Reindl, Analysis of passive RF-DC power rectification and harvesting wireless RF energy for micro-watt sensors, AIMS Energy 3 (2) (2015) 184−200.
[67] M.K. Hosain, A.Z. Kouzani, S. Tye, A. Kaynak, M. Berk, RF rectifiers for EMpower harvesting in a deep brain stimulating device, Australasian Physical & Engineering Sciences in Medicine 38 (1) (2015) 157−172.
[68] K.N. Puniran, R. Ahmad, R.A. Dziyauddiu, RF energy harvesting with multiple sources in wireless sensor network, in: IEEE 4th International Conference on Smart Instrumentation, Measurement and Application (ICSIMA), 2017, p. 1.
[69] F. Xie, G.-M. Yang, W. Geyi, Optimal design of an antenna array for energy harvesting, IEEE Antennas and Wireless Propagation Letters 12 (2013) 155−158.
[70] F.S. Mohd Noor, Z. Zakaria, H. Lago, M.A. Meor Said, Dual-band aperture-coupled rectenna for radiofrequency energy harvesting, International Journal of RF and Microwave Computer-Aided Engineering 29 (1) (2019) e21651.
[71] I. Buchmann, Batteries in a Portable World: A Handbook on Rechargeable Batteries for Non-engineers, second ed., Cadex Electronics, Inc., 2001.
[72] F. Fei, J.D. Mai, W.J. Li, A wind-flutter energy converter for powering wireless sensors, Sensors and Actuators A: Physical 173 (1) (2012) 163−171.
[73] N.S. Hudak, G.G. Amatucci, Small-scale energy harvesting through thermoelectric, vibration, and radiofrequency power conversion, Journal of Applied Physics 103 (10) (2008) 1−24.
[74] T.T. Le, Efficient Power Conversion Interface Circuits for Energy Harvesting Applications. Ph.D. thesis, Oregon State University, 2008.
[75] Y. Luo, L. Pu, Y. Zhao, RF Energy Harvesting Sensor Networks for the Healthcare of Animals: Opportunities and Challenges, 2018 (arXiv).
[76] P.K.K. Loh, S.H. Long, Y. Pan, An efficient and reliable routing protocol for wireless sensor networks, Proceedings of WoWMoM (2005) 512−516.
[77] M. Pedram, Design considerations for battery-powered electronics, in: Proceedings of Design Automation Conference, 1999, pp. 861−866. New Orleans, LA, June 21−25.
[78] S. Sudevalayam, P. Kulkarni, Energy harvesting sensor nodes: survey and implications, IEEE Communications Surveys Tutorials 13 (3) (2011) 443−461 (Third quarter).
[79] M. Zhu, E. Worthington, Design and testing of piezoelectric energy harvesting devices for generation of higher electric power for wireless sensor networks, in: Proceedings of the IEEE Sensors, 2009, pp. 699−702. Christchurch, New Zealand, October 25−28.

[80] W. Roh, J.Y. Seol, J. Park, B. Lee, J. Lee, Y. Kim, J. Cho, K. Cheun, F. Aryanfar, Millimeter-wave beamforming as an enabling technology for 5G cellular communications: theoretical feasibility and prototype results, IEEE Communications Magazine 52 (2014) 106−113.
[81] A. Nechibvute, A. Chawanda, N. Taruvinga, P. Luhanga, Radiofrequency energy harvesting sources, Acta Electrotechnical et Informatica 17 (2017) 19−27, 12.
[82] Q. Pham, F. Fang, H. Vu, M. Le, Z. Ding, L.B. Le, W. Hwang, A survey of multi-access edge computing in 5G and beyond: fundamentals, technology integration, and state-of-the-art, 2019 vol. Abs.1906.08452.
[83] H. Kong, I. Flint, P. Wang, D. Niyato, N. Privault, Exact performance analysis of ambient rf energy harvesting wireless sensor networks with ginibre point process, IEEE Journal on Selected Areas in Communications 34 (12) (2016) 3769−3784.
[84] S. Chen, F. Qin, B. Hu, X. Li, Z. Chen, User-centric ultra-dense networks for 5G: challenges, methodologies, and directions, IEEE Wireless Communications 23 (2016) 78−85.
[85] A. Kurs, A. Karalis, R. Moffatt, J. Joannopoulos, P. Fisher, M. Soljai, Wireless power transfer via strongly coupled magnetic resonances, Science 317 (5834) (2007) 83−86.
[86] J. Paradiso, T. Starner, Energy scavenging for mobile and wireless electronics, IEEE Pervasive Computing 4 (1) (2005) 18−27.
[87] X. Zhang, M. Jia, L. Chen, J. Ma, J. Qiu, Filtered-OFDM—an enabler for the flexible waveform in the 5th generation cellular networks, in: IEEE Global Communications Conference (GLOBECOM), Dec 2015, 2015, pp. 1−6.
[88] S. Divakaran, D.D. Krishna, N. Nasimuddin, Rf energy harvesting systems: an overview and design issues, International Journal of RF and Microwave Computer-Aided Engineering (2018) e21633.
[89] Y. He, X. Cheng, W. Peng, G. Stuber, A survey of energy harvesting communications: models and offline optimal policies, Communications Magazine, IEEE 53 (6) (2015) 79−85.
[90] S. El Hassani, H. El Hassani, N. Boutammachte, Rf energy harvesting for 5G: an overview, in: International Renewable and Sustainable Energy Conference (IRSEC), December 2017, 2017, pp. 1−6.

Innovations of the future of solar energy and COVID-19 impact

7

Dipti Shukla
O.P. Jindal University, Raigarh, Chhattisgarh, India

1. Introduction

Fig. 7.1 shows a natural source of energy (the Sun). Nothing on Earth is more essential to us than the Sun. The Earth would be a dead lump of ice-coated rock without the Sun's heat and light. The Sun heats our waters, stirs our atmosphere, creates weather patterns, and provides energy for the growing of green plants that support life on Earth with food and oxygen.

This implies that the Earth is in the proper location (neither too close nor too distant) to absorb the Sun's enormous energy, which includes the light and heat that is required for chemical processes. Because the Sun, like all stars, is essentially a huge fusion reaction, it can produce electricity. This process, known scientifically as nuclear fusion, produces a massive quantity of energy in the form of light and heat. However, there are

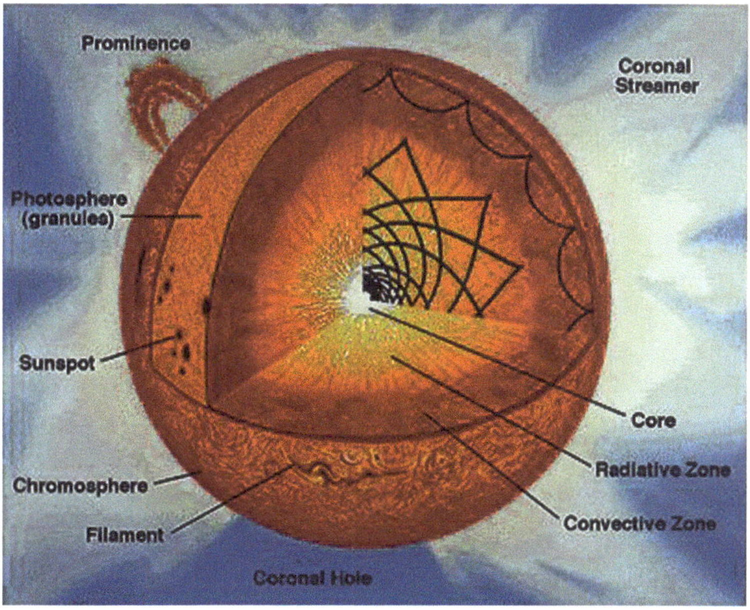

Figure 7.1 A natural source of energy (the Sun).

Green Energy Systems. https://doi.org/10.1016/B978-0-323-95108-1.00010-0

a few critical processes involved in transporting that energy from our Sun's core to Earth and beyond. In the end, it all comes down to the Sun's layers and the function each one plays in ensuring that solar energy reaches the places where it may contribute to the creation of new things. The Sun produces the equivalent of 384.6 septillion watts (3.8461026 W) at a mass–energy conversion rate of 4.26 million metric tonnes every second. This is the equivalent of around 9.1921010 megatons of TNT every second, or 1,820,000,000 Tsar Bombast—the most powerful thermonuclear weapon ever created. In Ref. [1], the author observed that every second, the Sun supplies around 1300 W per square meter to the Earth. Multiply the Earth's surface area and the number of seconds in a year. Solar energy is not a new concept. Solar energy has been employed since the 7th century BCE by humans. The Sun's energy has been respected and used in its most rudimentary form almost as long as man has roamed the planet. Solar power was first used to kindle fires for cooking by focusing on the Sun's energy through a magnifying glass. In the 3rd century BCE, the Greeks used the Sun to ignite holy torches for religious rites, and the Romans reflected sunlight off of "burning mirrors." While working with a cell comprised of metal electrodes in a conducting fluid in 1839, French physicist Edmond Becquerel discovered the photovoltaic (PV) phenomenon. He noticed that when the cell was exposed to light, it produced more power [2]. Solar cells may convert solar light directly into electricity (PV cells). When light contacts the junction between a metal and a semiconductor (such as silicon) or the junction between two separate semiconductors in such cells, a tiny electric potential is created, as shown in Fig. 7.2.

Other important occurrences in the history of solar energy include the solar panel in outer space and solar power aircraft. Solar panels, according to the National Renewable Energy Laboratory (NREL), are subject to typical wear and strain. Although there are no moving parts, chemicals can break down and degrade with time. Solar panels

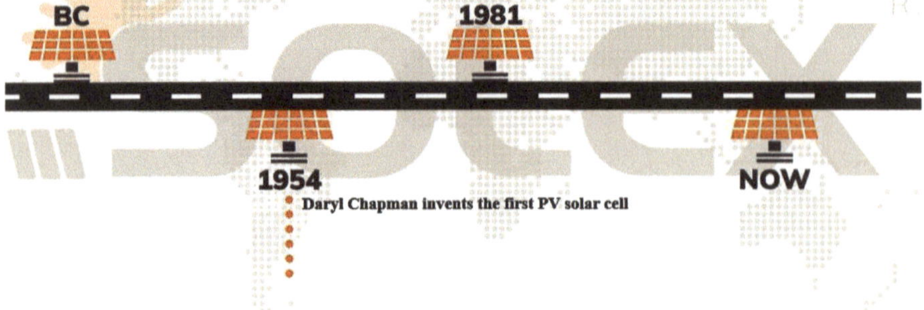

Figure 7.2 The history of solar energy.

from the 1970s and 1980s are still producing power, though not as much as they did previously. Solar panels have a very minimal risk of catching fire. In reality, according to *Photon* magazine, there has only been one incident every 10,000 installations. A house with correctly fitted solar panels, then, will not catch fire. To begin with, it should be emphasized that PV solar panels cannot create a fire on their own [3].

2. Solar panels generate electricity in a variety of ways

Fig. 7.3 shows electricity generation using solar panels.

Step 1: The panels are turned on by sunlight. A solar system with racks and panels.
Step 2: The cells generate electricity. An iron ore of silicon and a wafer of silicon.
Step 3: Electrical energy is turned into mechanical energy.
Step 4: Your home is powered by transformed electricity.
Step 5: A net meter is used to track use.

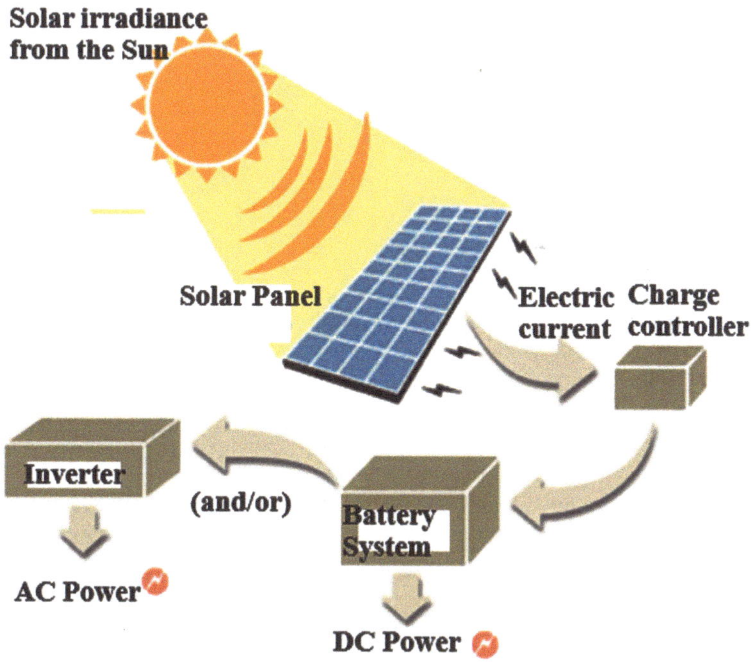

Figure 7.3 Solar energy—electricity generation.

3. Solar panel energy cost disadvantages

- The cost of acquiring a solar system is rather expensive at first.
- Dependent on the weather. Although solar energy may be collected during overcast and rainy days, the solar system's efficiency is reduced [4].

- It's expensive to store solar energy.
- It takes up a lot of room.
- Pollution is linked to this term.

Photovoltaic (PV) cells convert sunlight into energy and are used in solar panels. When solar panel waste is disposed of, it loses valuable resources. Land-filling also generates additional environmental dangers since solar panels contain harmful compounds like lead that can seep out as they degrade [5].

4. The future of solar energy

Solar will become even more affordable in the future as technology advances. Solar might very possibly become the most important source of energy for power production in most of the planet by 2030. This will also benefit the ecology and combat climate change. While most options have some advantages, they also have some disadvantages. Some of the disadvantages are as follows: alternative energy can be either expensive or inefficient, and storing it has proven challenging. Solar energy is a readily available, renewable source of energy. However, existing solar panels are inefficient. The most obvious disadvantage is that they are unable to generate power at night. While solar energy battery storage is increasing, this constraint implies that it may not be able to keep up with demand. The team is working on an "antisolar panel" to achieve this aim. Traditional solar panels employ PV technology to capture photons from the Sun's energy, which the system subsequently converts into electricity. Because they are colder than the Sun, they may absorb this energy. Heat will always go to a cooler place; therefore, it will flow through the solar panels [6].

Antisolar panel system: Fig. 7.4 shows antisolar panels that operate in the same basic way as solar panels, but in reverse. Because the air becomes cold at night, heat radiates forth from the Earth's surface, which has warmed during the day. This heat radiates in the form of infrared light, which these new features may use to create energy, a phenomenon known as radiative cooling. Antisolar panels have the potential to boost solar farm energy production.

Agrivoltaic: This approach was first proposed in 1981. In 2011, the term "agrivoltaic" was invented. Plants can thrive and produce behind elevated solar panels, and animals can graze the field under the panels, according to researchers. Solar energy in agriculture is now possible, and it has been shown to be effective. Furthermore, it provides a new source of revenue for farmers. Agrivoltaics is a fusion of solar panels and plants developed on farms that provides environmentally beneficial solutions. It simply refers to the employment of solar panels with food crops on the same piece of land to enhance land productivity. It's a concept that has the potential to take food production to the next level. In a study area in Montpellier, France, it was revealed that agrivoltaic systems are actually quite efficient: worldwide land production may improve by 35%–73%. revealed that the productivity of land might be increased by 60% as

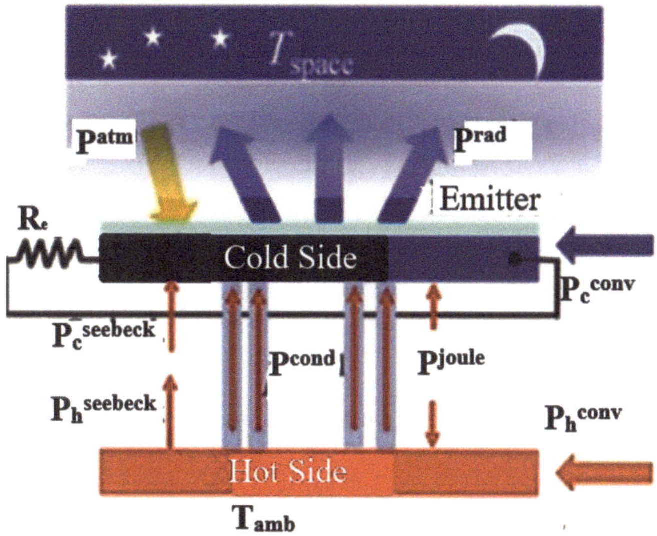

Figure 7.4 Antisolar panel.

shown in. The project results from the first year have been a great success: the agro-photovoltaic system shown in Fig. 7.5 has been proven to be adequate for practice and costs about the same as a conventional system.

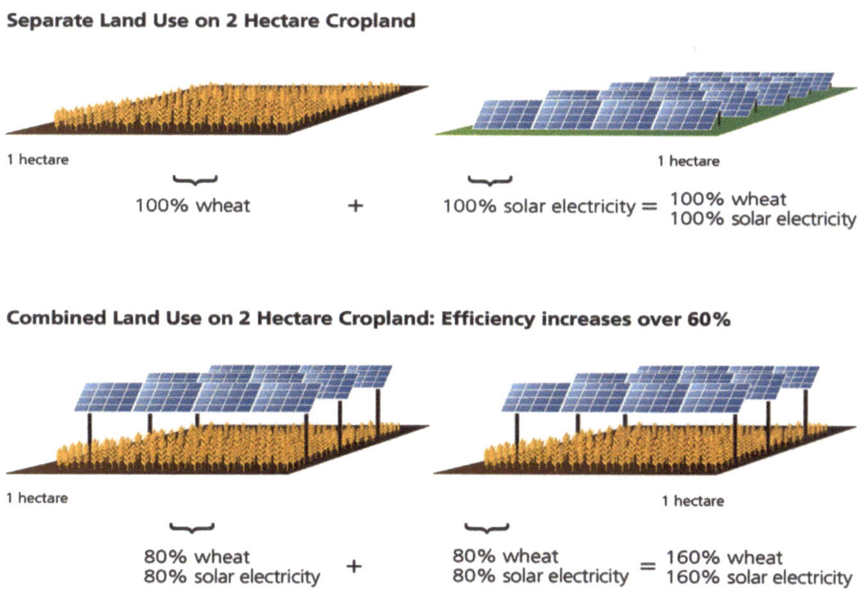

Figure 7.5 Environmental benefits through using agrivoltaics.

Advantages of Agrivoltaics:

- Farmer impact: Most people would assume that solar panels occupying farmland is a bad thing. However, farmers benefit from agrivoltaics technology because they can grow crops and earn money from solar generation on the same piece of land.

Environment impact: Agrivoltaics provides environmentally sustainable alternatives. This has an influence on how much food the land can currently produce. Peppers, broccoli, and Swiss chard, for example, can only yield approximately 60% of their volume when grown in full light. Furthermore, solar panels have roughly half the power generation capacity per acre and are more expensive. Solar power generating investment is rapidly increasing all over the world. Solar energy increased its share of worldwide power-producing capacity by 50% in 2016, surpassing wind, gas, and other renewable technologies in terms of growth. Since 2016 [7] it has been claimed that the cost of solar PV cells—the principal capital cost in solar projects that use that technology—has dropped by 80%. Agrivoltaics seem to be well adapted to market gardening in general, but not so well to arable crops. Agrivoltaic technology also lowers the maintenance concerns associated with solar panels that are placed closer together and allows the area to be used for agriculture. However, several concerns with cultivation operations must be considered, such as the size and effectiveness of farm equipment that may be deployed under and between the frames.

Floatovoltaics: Fig. 7.6 shows the general setup for floating PV solar power systems; they have been around in concept and design for a long time, yet they provoke curiously strong responses anytime they are mentioned. The fundamental benefit of floating PV plants is that they do not require any land, with the exception of the small areas required for the electric cabinet and grid connections. Their cost is equivalent to that of land-based plants, but they offer a smart option to avoid using dry land. Floating

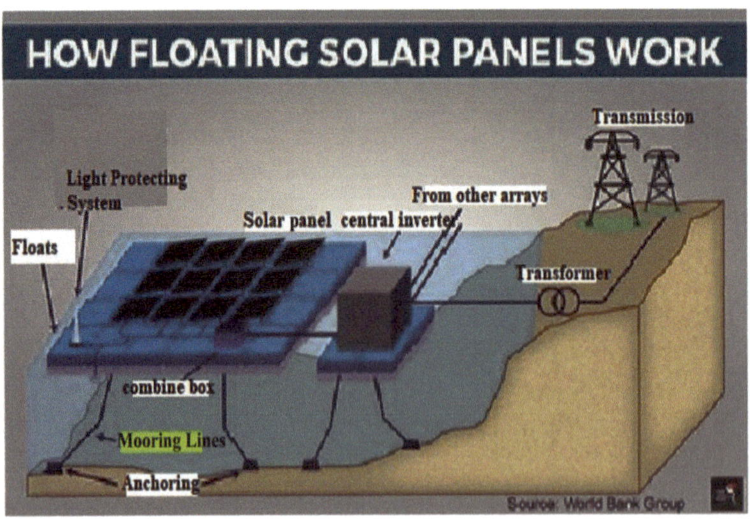

Figure 7.6 Floatovoltaics.

solar panels in floatovoltaics not only save space, they also conserve water and maintain its quality. In Ref. [8], the authors suggested that [2]. Floatovoltaics, also known as floating PV, is a new innovation for installing solar panels on a water body rather than on dry land. The solar panels are modular and are installed on floating platforms. They can either form a big structure or float as individual modules that keep their geometric layout. Because the surrounding water acts as a natural coolant for the panels, the efficiency of these floating structures is anticipated to be 8%−10% greater than that of land-based solar plants. This enhances the total energy yield while increasing efficiency [9].

5. Advantages of floatovoltaics

- Increased energy conversion, improved water quality, less evaporation, decreased panel soiling, and less civilian infrastructure are just a few of the benefits of installing a floatovoltaic system. The benefits of the floatovoltaics technology are described further [10].

The PV system's energy efficiency has improved. The efficiency of a PV system is determined by a number of factors. The major difficulty is its dependency on temperature. The efficiency of PV modules decreases as the temperature of the panel rises. Floating PV systems are equipped with a cooling mechanism that makes use of the water in which they float as a natural heat sink. As a result, they are more efficient than comparable ground-based systems [10].

Enhances the water quality: Photovoltaic systems save both water and productive land for future generations. Photosynthesis allows organic matter, including algae, to thrive when water bodies are exposed to light. These algae are typically undesirable because they can clog pumping and filtering systems, necessitating expensive chemical treatment to eliminate the problem. The placement of floating solar panels decreases photosynthesis by shading the water. Algal blooms are less likely to occur, and chemical and operating expenses are lower as a result [11].

Evaporation is reduced: Floatovoltaics minimizes water evaporation from a water body by reducing the reservoir's effective surface area, which is a substantial advantage for nations with limited water supplies. On natural lakes and ponds, floating solar panels can reduce water evaporation by up to 33%, and on man-made facilities by up to 50% [12], Zion Market Research.

Panel soiling is reduced: The installation prevents soiling since the solar panels are situated on the water's surface, away from sources of dust and dirt. The distance from the ground, especially in dry and dusty places, prevents dust and grime from being captured, decreasing the need for surface cleaning. Furthermore, a large amount of water is used to clean the surface of such facilities.

There will be less civil infrastructure work. Because floating systems do not require thousands of metal frames to be anchored to the ground, a panel array may be built faster. Furthermore, decommissioning a floating system is significantly easier and less costly.

Solar window: Solar windows are essentially windows with solar panels incorporated into them. They appear like regular windows, but they have PV glass that turns sunlight into renewable energy. This implies that households who want to create their own renewable energy do not require solar panels to do it. Solar cells are embedded in solar glass windows, which transform sunlight into useful energy. This shows that throughout the day, light streaming through the windows might provide free renewable power for a household. Any extra energy will be transferred to the national grid or, if one is available, stored in a solar battery for later use. Using free renewable solar energy to generate electricity is a cost-effective approach to powering a house. Solar panels, on the other hand, are not appropriate for every situation. More individuals are able to create their own free renewable power thanks to solar windows. Solar windows provide more than just a free source of electricity that may help save money on energy costs, as shown in Fig. 7.7.

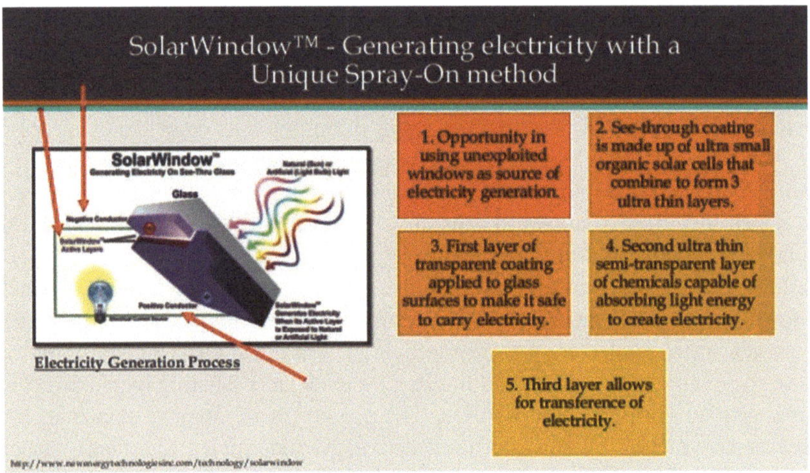

Figure 7.7 Solar window electricity generation with a unique spray-on method.

6. Advantages of solar windows

UV rays should be avoided: UV radiation from the Sun can cause fading and other damage to furniture. Solar windows help to keep as much UV radiation out of the house as possible, allowing your furnishings to last longer.

Increased personal privacy: Solar windows feature a tint that makes it difficult to look into the building. When looking out the window from the inside, you will be able to see just as well as before.

Glare reduction: Glare may be aggravating, especially when it's on a television. Glare may be significantly decreased with the use of solar windows.

Solar cell fabric: Fig. 7.8 shows solar cell fabric, which is a fabric having PV cells incorporated in it that produce energy when exposed to light. Solar cells made of

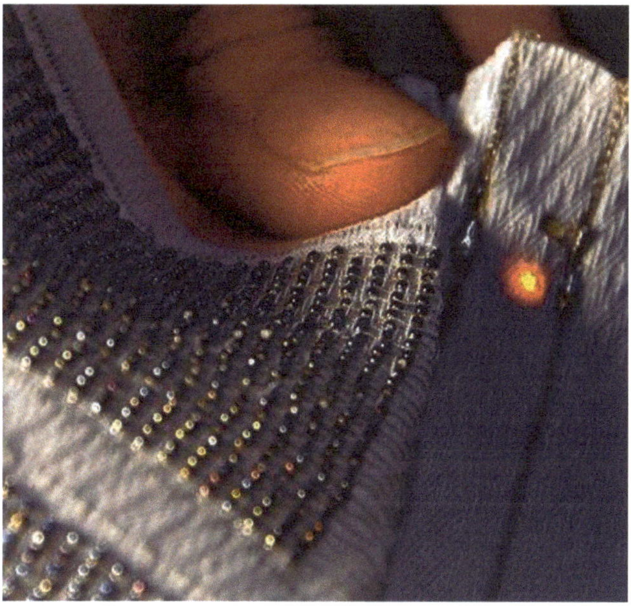

Figure 7.8 Solar cell fabric.

traditional silicon are expensive to produce, inflexible, and brittle. Thin-film cells and organic polymer-based cells can be made fast and cheaply. However, they are less efficient. They are also stretchy and can be sewn onto cloth. Unlike traditional panels, solar fabric can be bent or bonded to any surface, is 10 times lighter than framed panels, and is free of harmful ingredients. These also have a finite lifetime, lasting up to 20 years. The solar fabric captures less energy than standard solar panels (13% vs. 15%–18%), but it is more stable when temperatures vary, operating even on cloudy days by capturing a wider spectrum of light. Some solar panels are made of glass or other composite material that can resist the elements, especially in harsh environments like deserts. However, the combination of metal and glass is unsuitable for clothes. As a result, solar-powered textiles are becoming increasingly popular.

7. Advantages of a solar fabric system

Now that you have a better understanding of the solar cell fabric, let us look at some of the advantages it provides:

- It aids in the absorption of solar energy.
- It permits solar energy to be converted into electrical energy.
- It is favorable to the environment.
- Solar cell fabric may be used on a variety of surfaces.
- It can be cost-effective since it can help to save money by reducing the amount of power consumed.

8. Solar cell fabric's working mechanism

The concept of operating a solar cell fabric is quite simple.

When light strikes a solar cell, electrons in the absorber layer are stimulated from a lower energy or ground state to a higher energy or excited state. They are attached to specific atoms in the solid at this step, allowing them to pass through the solid in a higher or excited state. The unbound electrons are in random motion if the junction-forming layers are missing. As a result, there cannot be any direct current (DC). When junction-forming layers are present, however, they generate an electric field that generates the PV effect. The electrons that pass across the electrical contact layers to an external circuit experience a collective motion as a result of the electric field. The PV process is analogous to photosynthesis in some ways. Solar cells have backup energy saved in many applications since they do not operate in the dark. Charging electrochemical storage batteries is the most popular method of conserving electrical energy. You can use the solar unit for a long time if you do this, even if there is no sunshine.

9. Solar cell fabric's future

Solar cell textiles are now in the experimental stage. According to researchers, these materials may generate energy by inserting cell textiles into garments, curtains, and tents. As a result, solar fabric apparel and solar fabric items have become available. Such things can be used for a variety of reasons. However, PV goods are not currently available on the market. In the interim, a portable and simple PV gadget can keep your smart phone powered.

Although solar cell fabric is not currently available, inexpensive household and commercial PV systems exist.

Solar thermal fuel: Fig. 7.9 shows a solar thermal collector. Solar thermal fuel is a chemical that can absorb sunlight, store it for several days, and then release it as heat. Despite the fact that this fuel is not yet an established technology, it offers a wide range of heating uses. A substance that can absorb solar energy, store it for a longer length of time, and then release it is known as solar thermal fuel. Solar thermal fuel stores solar energy in the form of heat through light-induced changes in its chemical structure and releases it as needed. Solar thermal fuel technology was invented in 2018 by Chalmers University of Technology (Sweden) experts as a specialized fluid that can supposedly store solar energy for up to 18 years. It works like a rechargeable battery, except that, instead of electricity, it uses sunshine to charge and discharge heat. The fluid has been tested for more than 125 cycles and may be reused in the process. A group of Chalmers University researchers began building prototypes of solar thermal fuel technology for large-scale use around 2020. When sunshine is not available, a solar fuel may be created and stored for later use, making it a viable alternative to fossil fuels and batteries. Hydrogen, ammonia, and hydrazine are examples of such fuels. Solar fuel technologies are experimental systems that create fuels that can be utilized in our current

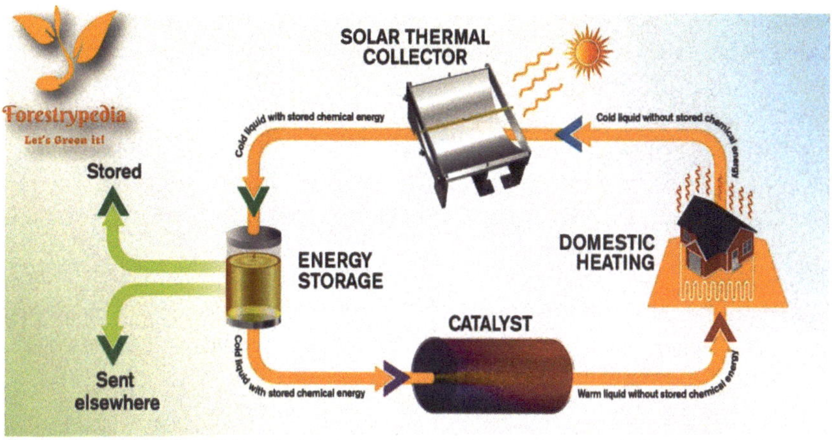

The energy system MOST works in a circular manner—completely free of emissions, and without damaging the molecules carrying the energy.

Figure 7.9 Solar thermal collector.

energy infrastructure by combining resilient, low-cost, and high-efficiency components. Solar fuel technologies create fuels from sunshine, water, carbon dioxide, and nitrogen in the air. Solar thermal electric energy generation focuses on the Sun's light to produce heat, which is then utilized to power a heat engine that spins a generator to generate electricity. All of these engines are capable of producing tens to hundreds of megawatts of electricity and may be highly efficient, averaging between 30% and 40%.

10. Impact of COVID-19 on the global solar thermal fuel market

The worldwide solar thermal fuel market has been harmed by the COVID-19 outbreak. Various research and development laboratories in various sectors have temporarily shut down due to the ongoing lockdowns. The crisis has had a severe impact on alternative technology research, since both the corporate and public sectors have refrained from investing in the area. However, favorable government assistance packages, which will be enacted following the lockdowns, are anticipated to result in a rapid rebound of the solar thermal fuel industry in the near future.

11. Vision for solar energy's post-COVID-19 future

The COVID-19 pandemic has hastened the use of renewable energy in the future, according to a poll the following results were found.

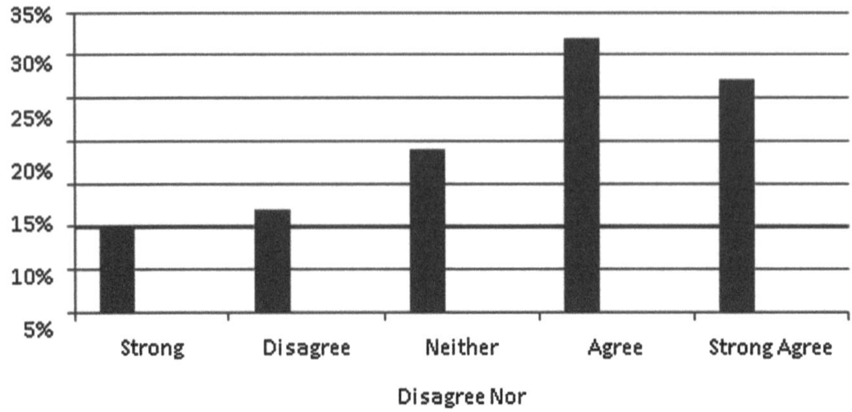

Figure 7.10 Poll percentage about the COVID-19 pandemic has hastened the use of renewable energy in the future.

COVID-19 saw a considerable drop in fossil fuel-based energy output, while renewable energy gained traction. In a poll carries out by Verdict to determine if COVID-19 functioned as a catalyst for carbon reduction by opening the way for future renewable energy use, a majority of 59% said it did, with 27% strongly agreeing and 32% agreeing, as shown in Fig. 7.10.

On the contrary, 22% of respondents opposed or strongly disagreed with the idea that COVID-19 served as a catalyst, with 12% disagreeing and 10% severely disagreeing.

Furthermore, 19% of respondents were undecided on whether COVID-19 functioned as a catalyst for carbon reduction by laying the way for future renewable energy use. Between November 18, 2020, and March 17, 2021, 248 replies were received from readers of Power Technology, a Verdict network site.

12. COVID-19's impact on future renewable energy use

The COVID-19 pandemic-related lockdown limitations resulted in an 8% reduction in energy-related emissions and a 6% reduction in energy consumption, resulting in an increase in renewables' share of power output. During the first 10 weeks of the lockdown, renewable energy usage in the United States climbed by 40% and in India by 45%. Global renewable energy generation increased by 3%, owing to the addition of new solar and wind projects in recent years. Despite the epidemic, 13 nations were granted additional renewable capacity additions totaling almost 50 GW in the first half of 2020.

COVID-19 depicts a future powered by wind, however solar energy has a battery shortage problem.

Coronavirus has shown a dearth of investment in large batteries, which are essential for harnessing solar and wind power. Due to the pandemic's decline in energy consumption, European networks are now overwhelmed with green electricity, posing a threat of blackouts and emphasizing the importance of energy storage in a low-carbon energy system. The worldwide COVID-19 pandemic is altering the solar landscape in ways that have never been seen before.

On the one hand, it has awakened nations like India to the need to develop their own supply chains, but it has also driven home the importance of a shared energy future. Dharmendra Pradhan, India's Oil Minister, addressed delegates at the International Solar Alliance's (ISA) First World Solar Technology Summit, "We are in the process of reforming India's supply chain and reducing overdependence on imports of solar modules." In addition, India's Ministry of New and Renewable Energy, the World Bank, and the International Solar Alliance signed a tripartite agreement to construct a "One Sun, One World, One Grid" initiative with the goal of interconnecting grid networks so that member nations are able to provide each other with power at peak hours.

COVID-19: A postpandemic approach to solar power project execution and operation and maintenance: It is possible that business goals may shift in the post-COVID-19 future, but the necessity for better execution and O&M will remain the same. Renewable energy has become important in practically all nations throughout the world, including India, during the last decade or so. Solar power is expected to account for the largest share of the future energy mix.

According to data provided by India's Ministry of New and Renewable Energy, the country has roughly 32.53 GW of installed solar power capacity, with another 25.05 GW under construction and another 25.78 GW being bid. When the COVID-19 outbreak occurred, India's progress toward its solar energy goals took a hit.

There was undoubtedly a short-term impact, but there were some benefits as well.

Solar developers and solar equipment manufacturers in the country are concerned about interruptions in global and domestic supply chains, at least in the short-to-medium term. Furthermore, the pandemic has hampered the progress of ongoing initiatives. According to CRISIL, delays might stymie projects worth Rs 160 billion, totaling 3 GW of solar generating potential. According to certain projections, India may only be able to install around half of the planned solar power capacity in the first quarter of 2020. However, this short-term impact pales in comparison to the bleak picture for the following several years, which is anticipated to be marked by decreased electricity demand.

The key to successful execution is cost-effectiveness and resource allocation.

Solar projects in India are typically built utilizing either the EPC (engineering, procurement, and construction) or package models. If there are not as many experienced resources as are required to handle the project, the former is preferred. The package approach, on the other hand, enables businesses to deploy huge teams for projects at a reasonable cost.

Any plant design optimization that is required must be done in accordance with the capital expenditure specified in the winning offer. Optimizing plant capacity and utilizing private investments to reduce capital costs can aid in the creation of low-cost generating assets. If the project entails the building of a transmission line or a substation, it's a good idea to have a specialized team in place from the start to look into right-of-way concerns.

References

[1] P. Denholm, The technical potential of solar water heating to reduce fossil fuel use and greenhouse gas emissions in the United States (PDF), 2007, pp. 12–28.
[2] D.M. Chapin, C.S. Fuller, G.L. Pearson, A new silicon p-n junction photocell for converting solar radiation into electrical power, Journal of Applied Physics (6) (1954).
[3] D. Roberts, California will require solar panels on all new homes. That's not necessarily a good thing, 2018.
[4] K. Bódis, I. Kougias, N. Taylor, A. Jäger-Waldau, Solar photovoltaic electricity generation: a lifeline for the European coal regions in transition, Sustainability 11 (13) (2019) 3703, https://doi.org/10.3390/su11133703.
[5] M.A. Green, Photovoltaic technology and visions for the future, Progress in Energy (2019) 013001, https://doi.org/10.1088/2516-1083/ab0fa8.
[6] K. Shravan, K. Chunduri, PERC Solar Cell Technology-2018 Edition, 2018.
[7] H. Dinesh, J.M. Pearce, The potential of agrivoltaic systems, Renewable and Sustainable Energy Reviews 54 (2016) 299–308, https://doi.org/10.1016/j.rser.2015.10.024.
[8] N.M. Kumar, E. Mallikarjun, E. Engineering, Floatovoltaics: towards improved energy efficiency 9, 2018, pp. 1089–1096.
[9] K. Thoubborn, Floating solar: what you need to know, 2018, pp. 1723–1724.
[10] M. Willuhn, Tracking the wind, PV magazine photovoltaic markets and technology, 2019.
[11] C. Roselund, Bifacial module demand, PV Magazine Photovoltaic markets and technology, 2019.
[12] B. Norton, Harnessing Solar Heat, 2013.

Potential of ionic liquids in green energy resources

8

Nidhi [1], *Ramesh Chandra Thakur* [2], *Shimaa M. Elsaeed* [3] *and Ashish Kumar* [1,4]
[1]Department of Chemistry, Faculty of Technology and Sciences, Lovely Professional University, Phagwara, Punjab, India; [2]Department of Chemistry, Himanchal Pradesh University, Shimla, Himachal Pradesh, India; [3]Petroleum Applications Division, Egyptian Petroleum Research Institute (EPRI), Cairo, Egypt; [4]NCE, Department of Science and Technology, Government of Bihar, Patna, India

1. Introduction

Green energy, also called green power, is the energy generated from natural resources like solar, geothermal, wind, biomass, or water, providing a huge environmental benefit. Using green energy is consuming a mix of green, renewable, as well as conventional energy [1]. As we discuss green energy we have biomass, solar energy, hydroelectric energy, wind energy, and a lot of other kinds of sustainable energies [2]. And among all these energies, biomass which was used even before the era of fossil fuel is considered as a very essential and efficient alternative. And after biomass, solar energy is said to be the future energy because of its utilization for photothermal, photochemical, photoelectric, and photobiological conversions. These energies are followed by other energies such as hydroelectric or geothermal energy with different applications in various fields [3].

In our current situation, to enable us to fulfill all the energy demands of the population all these energies produced from renewable sources are predominantly found to be efficient alternatives. As the demand for the energy produced from fossil fuels is increasing it is rapidly exhausting these fossil fuels as well as creating environmental issues like global warming and pollution. Therefore, to fulfill the requirements as well as to protect the environment from the degradation and scarcity of resources, the United Nations has also set the SDG-7 which aims to lower the harmful impacts of fossil fuel utilization by making renewable resources accessible to everyone by 2030. And, along with these rules, the production of renewable energy via green processes, because of their biodegradable, easy availability, cheap, and nontoxic nature, is in the limelight [4]. Also, ionic liquids (ILs), being from the family of novel media with characteristics like polarity, solvent miscibility, high thermochemical stability, and a lot more, have been considered as an alternative for the production of green energy in various forms.

Discussion about ILs has gained significant interest in recent years due to their specific set of properties that make them useful in many fields.

Green Energy Systems. https://doi.org/10.1016/B978-0-323-95108-1.00005-7

ILs is basically a generic term for a class of salts in molten/liquid state with a melting point equal to or below the ambient temperature, such as 100°C. ILs are the composition of ions or the interaction of cations and anions. These are also described as ionic melts, fluids, liquid salts, and liquid electrolytes [5].

ILs are receiving increasing interest as the most promising alternative because of their unique combination of properties such as low melting point, high ionic conductivity, high ion density, negligible volatility, high polarity, and high thermal and chemical stability. These properties are mainly due to their composition as well as the strong ionic bond between the ions, which is much greater than the van der Waals forces [6].

Also, due to these great properties ILs are believed to be important in a number of energy-related applications and have been utilized as catalysts or solvents in clean industrial technology with significant cost and also a lot of environmental benefits.

On the basis of the temperature range, ILs may be classified in categories such as low-temperature ILs and room-temperature ILs (RTILs) [7]. Low-temperature ILs are comparable to ionic solutions with ions as well as neutral molecules. And due to this composition, these kinds of ILs are also described as mixtures of ionic and nonionic substances which possess a very low melting point which is very much lower than that of pure liquids.

On the other hand, RTILs are composed of bulky and asymmetric organic and inorganic cations and anions. Cations may be ammonium ions, 1-alkyl pyridinium, 1-alkyl-3-methylimidazolium, or phosphonium ions, and anions like tetra fluoroborate, triflate, and halides. Generally, the melting point of halides is higher than that of other inorganic solvents [8].

These ILs have a very weak ionic interaction, i.e., the bonds between the ions of these types of ILs are not strong and can be broken easily, even at room temperature. Therefore, they exist in liquid state at room temperature and called room-temperature ILs [9].

Due to their scientific and technological importance, both low-temperature ILs and room-temperature ILs have their own specific properties in different fields. For example, low-temperature ILs are advantageous in imaging long-wave infrared light and also to prevent evaporation in lunar vacuums such as RTILs because of their low volatile and heat-resistant nature, mainly used in electrochemical processes, batteries, fuel cells, electrocatalysis, and electrochemical biosensing [10,11].

The unique combination of cations and anions in ILs provides RTILs with the tunable ability to be used as potential energy storage materials.

ILs are multiple-purpose materials due to the wide range of possible binary and tertiary ILs as they offer a wide range of solvents and also replacements for conventional media or chemical processes.

The different combinations of cations and anions for ILs could also be investigated for use in energy storage devices. Of all the kinds of ILs the most suitable as well as those most popular currently are triazolium- and pyrrolidinium-based ILs.

These liquids are considered advantageous in energy storage applications not only because of their unique versatility but also because of their green credentials, high solubility for organic and inorganic substances, high stability, and low toxicity [10,12].

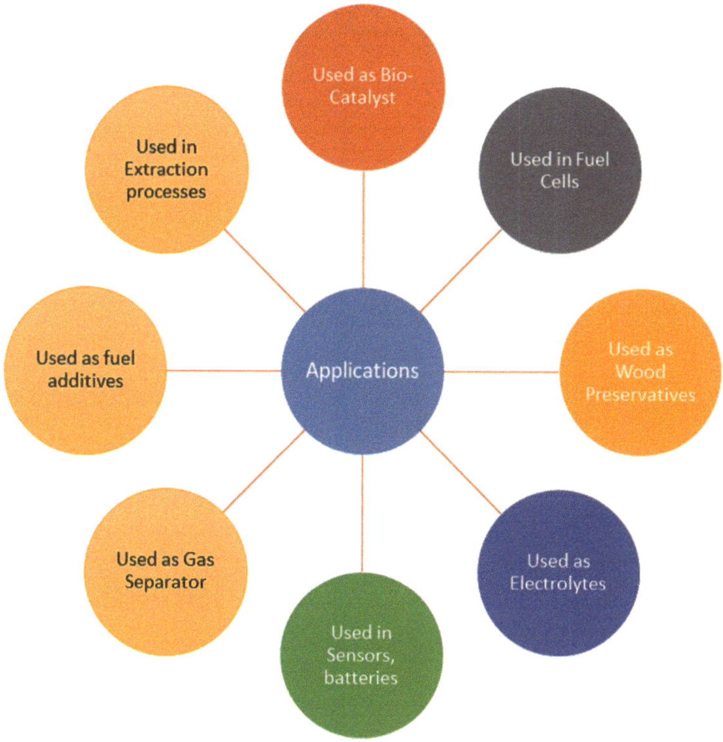

Figure 8.1 Applications of ionic liquids.

Because of their great solubility toward organic and inorganic materials, for our studies in enhancement of energy storage applications we have taken amines along with these ILs and investigate their properties as a combination and analyze their behavior in enhancing the applications of energy storage applications [13–15]. Fig. 8.1 provides examples of applications of ILs as an electrolyte, catalysts, additives, separators, wood preservatives, etc.

2. Applications of ionic liquids

ILs are formed by the interaction of cations and anions. These interactions of different cations and anions give the special character, or specific characteristics, to these liquids. Table 8.1 provides a list of different cations and anions used in electrochemistry for the synthesis of the ionic liquids.

ILs are also known as molten salts which are composed of organic cations with characteristic features such as ion conducting, capacitance, wide electrochemical window, nonflammability, etc. They are also known as green solvent because they are considered as safe for the environment. ILs are commercially viable for energy storage applications and have been used in many devices for enhancement of their performance, including fuel cells, supercapacitors, batteries, and dye-sensitized solar cells.

Table 8.1 List of cations and anions used in ILs.

S. No.	Ionic liquid	Structure
1	R,R′-imidazolium	$[R_2 - Im - R_1]+$
2	R,R′-pyrrolidinium	$[R_2 - pyr - R_1]+$
3	Dicyanamide	PF_6^-
4	p-Toluenesulfonate	$C_7H_7 - SO_3^-$
5	Bis(trifluoromethanesulfonylamide)hexafluorophosphate	$(CF_3SO_2)_2N^-$

A large number of ILs have been investigated including phosphonium-, pyrrolidinium-, pyridinium-, imidazolium-, and triazolium-based ILs. Many different compounds can be created by using a different combination of anions and cations with different and specific sets of characteristics serving different applications in numerous fields [10,16–19]. Some of the categories of ILs have already been used in many fields and show a great purpose in green energy resources. Some of these ILs, along with their chemical depiction, properties, and applications are explained further below.

2.1 Imidazolium-based ionic liquids

These are the versatile ILs, also known as green organic solvent. Imidazolium-based ILs show a unique combination of alkyl substituents, allowing a specific set of characteristics. The best thing about these ILs is their tunable structure along with other specific properties including high ionic conductivity, high electrochemical window, and amphoteric nature. These properties of this type of IL have drawn the attention of many fields including energy storage applications, making them more environment friendly such as 1,3-dialkylimidazolium which has been studied intensively. As these ILs have high conductivity as well as great electrochemical stability, they have great importance in energy storage devices as they are used to increase the stability of Li-salts in Li-ion batteries, along with salt stability they also increase the yield and chemoselectivity when used as a catalyst [20].

These ILs are further categorized into four categories:

1. Monosubstituted
2. Disubstituted
3. Trisubstituted
4. Benzimidazole.

Fig. 8.2 shows different categories of the imidazolium-based ILs along with their structures.

Fig. 8.3 includes some of the imidazolium-based ILs along with their structure which has been investigated and are suitable for application in electrochemistry.

S.No	Ionic Liquid	Structure
1	1-n-butyl-3-methylimidazoliumfluoromethanesulfonate	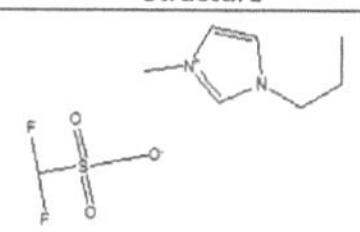
2	1-n-butyl-3-methylimidazoliummethanebromide	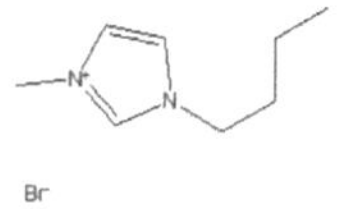
3	1-n-butyl-3-methylimidazoliumchloride	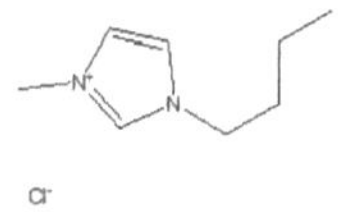
4	1-n-butyl-3-methylimidazoliumhexafluorophosphate	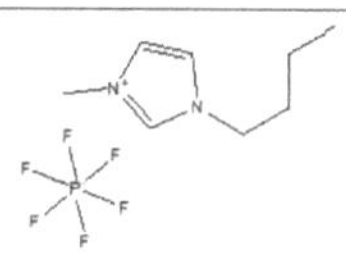
5	1-n-butyl-3-methylimidazoliumtetrafluoroborate	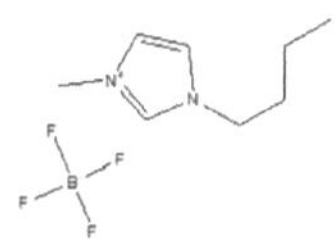

Figure 8.2 Different categories of imidazolium-based ILs.

2.2 Ammonium-based ionic liquid

Ethanolammonium nitrate was the first ammonium-based IL, and was synthesized in 1888 by Gabriel.

Ammonium-based ILs are water and air sensitive with low viscosity, great cathodic stability, and a low melting point; their almost negligible viscosity along with high electrochemical stability makes them suitable for electrolytes in high-energy electrochemical devices. Specifically discussing energy storage applications, cyclic alkyl quaternary ammonium (QA) cations are the class of cations whose room temperature along with high thermal and electrochemical stability that has been very useful in energy storage applications [21]. Fig. 8.4 provides us a list of some ammonium-based ILs which have been investigated for energy storage applications.

2.3 Sulfonium-based ionic liquid

These ILs have low viscosity, high conductivity, and an electrochemical window. In these types of ILs, molecular dynamics cannot be carried out because of their properties such as their hydrophobicity and high chemical stability against moisture. Also, these are considered as appropriate for apolar molecular liquids. However, for these kinds of ILs the molecular dynamics simulations are done to describe them at an

S.No.	Ionic Liquid	Structure
1	Monosubstituted	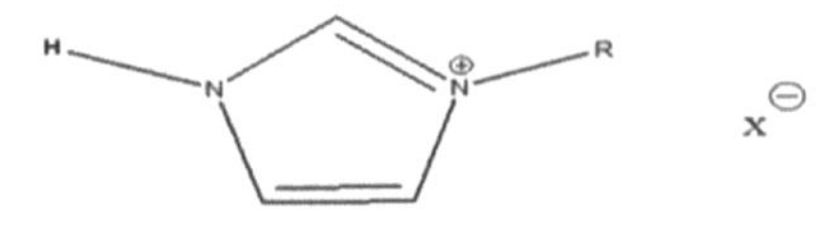
2	Disubstituted	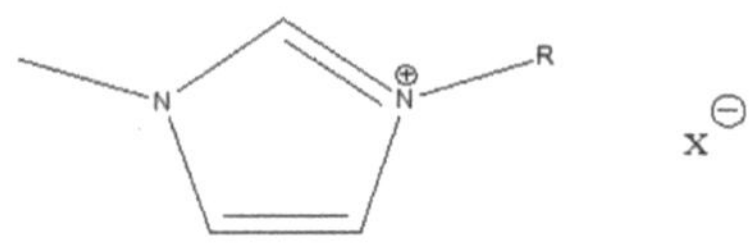
3	Trisubstituted	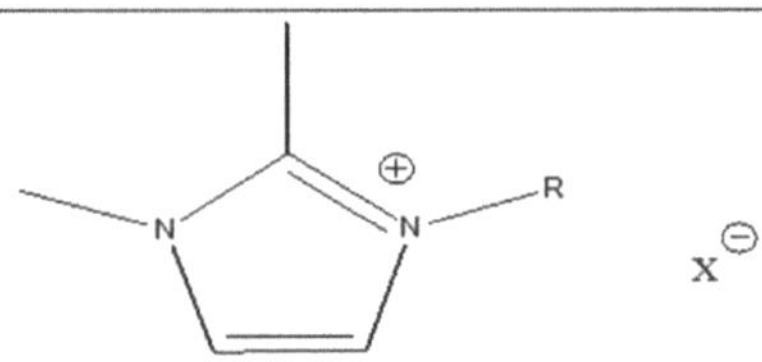
4	Benzimidazole	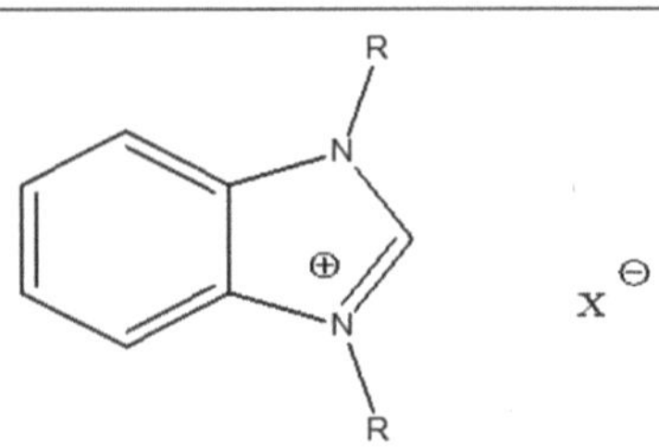

Figure 8.3 Common imidazolium-based ILs.

atomic level for the dynamical and electrical properties of some of the selected sulfonium cations which are used in supercapacitors.

Due to their thermochemical stability, these ILs are used as potential alternatives for their ammonium counterparts. Among all the sulfonium-based ILs bis(trifluoromethylsulfonyl)imide-based ILs and $[NTf^2]^-$ and bis(trifluoromethylsulfonyl)imide-based ILs have been investigated and concluded as the best alternatives to be used in energy applications like Li-ion batteries, solar cells, or electric double-layer capacitors or supercapacitors. Fig. 8.5 lists some of the sulfonium-based ILs which have been investigated and considered as safe for the environment [22].

3. Ionic liquids in green energy

While discussing the potential of ILs in green energy, the first thing which comes to mind is the production of green energy or the production of fuels using green products.

S.No.	Ionic Liquid	Structure
1	**Tributyl methylammonium chloride**	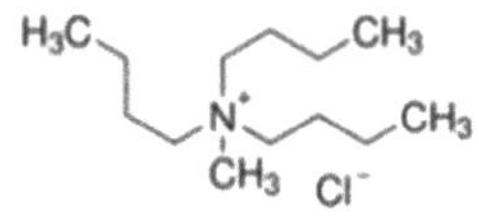
2	**Butyl trimethylammonium bromide**	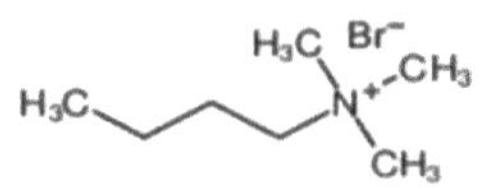
3	**Tribuytl(methyl)ammonium (Methyl tributyl ammonium chloride)**	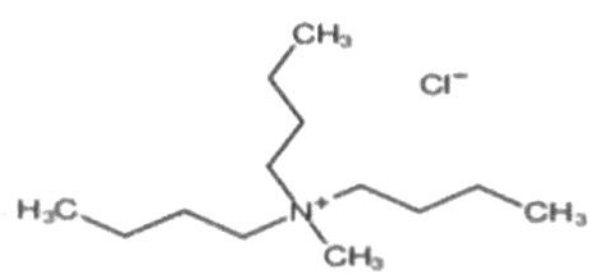

Figure 8.4 Some ammonium-based ILs.

S.No.	Ionic Liquid	Structure
1	**Diethylmethylsulfonium bromide**	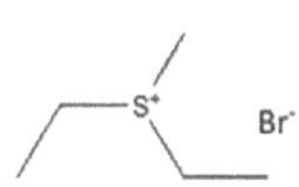
2	**Trimethylsulfonium bromide**	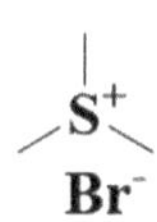
3	**Bis(trifluoromethanesulfonyl)amide**	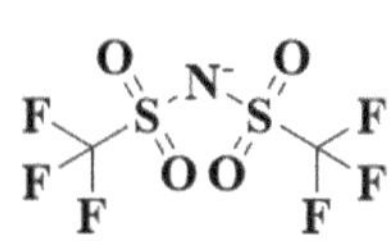

Figure 8.5 Common sulfonium-based ILs.

ILs, because of their specific set of characteristics, such as negligible vapor pressure, thermal stability, transport properties, and great solubility toward the organic solvents, have made a great impact on the production of some fuels or fuel components from different green components [23–25]. Some of the conversions which are catalyzed using ILs are discussed in the following sections.

3.1 Conversion of CO_2 into fuel or fuel components

In our environment, the most challenging issue we are facing is control of the concomitant consumption of greenhouse gases. To get this usage under control we need the conversion of greenhouse gases into energy, mainly carbon dioxide because it is a safe and renewable carbon source. The conversion of CO_2 into energy or fuel or fuel additives not only gives us an alternative for energy but also provides us with energy that can be recycled [26].

The conversion of carbondioxide into energy is not an easy task because of its high oxidation state along with the nature of inertia. Therefore, to make this conversion easy and possible, ILs, because of their specific properties of solubility enhancement and electrochemical catalysis of CO_2, have provided a breakthrough in this field. We can categorize this conversion into the following groups in accordance with their different fuel routes [27].

3.2 Conversion of CO_2

3.2.1 Carbon monoxide

Carbondioxide can be reduced to carbon monoxide, which can be further used by transforming it into many products such as in synthesis of oil and many more; this $CO_2 \rightarrow CO$ conversion is considered to be the most important technology for the proper usage of CO_2 because of its further use. The techniques used to reduce CO_2 into CO reverse the water gas shift reaction and photochemical and electrochemical reactions under mild conditions are useful [28].

Also, to enhance some of these technologies, ILs have been used as a catalyst in the reaction, especially the reverse water gas shift reaction. Using this reaction, Tominaga et al. [29] composed a system of an ionic Ru cluster in the presence of ILs as a catalyst, i.e., bis(triphenylphosphine)iminium chloride ([PPN]Cl), at a temperature range of 160°C. This hydrogenation process of the hydride cluster is followed by CO_2 coordination and the proton's electrophilic attack on its oxygen atom in the presence of chloride anions [30].

3.2.2 Formic acid

We can produce formic acid by two methods, and of these two one is the experimental method for electricity storage where we can produce formic acid from sunlight and the other is by the hydrogenation of carbon dioxide in polar solvents. With the use of nitrogenous bases, the other method for the same conversions is the hydrogenation of carbondioxide using an integrated sCO_2 process with mobilized catalyst and base to avoid the formation of formate salts causing pollution and other complexities [31–33].

To make this conversion environment friendly, we used ILs in the production of formic acid. This plays an important role in the formation of formate salt via neutralization and its result shows that in one reaction cycle the addition of ILs facilitated 2 moles of formic acid per mole of IL.

3.2.3 Alcohols

Methanol can directly be used as a fuel or fuel additive and has been used in applications in various industrial phases. Therefore, using methanol as a base chemical for synthesizing various fuels, biodiesel has been considered and worked on.

$$CO_2 + 3H_2 \rightarrow CH_3OH + H_2O \tag{8.1}$$

$$CO_2 + 2H_2 \rightarrow CH_3OH$$

The above equations represent the reaction which tells us that methanol can directly be produce fromd CO_2/CO or H_2 using these reactions. Here also, ILs like imidazolium-based ILs play an important role as a catalyst to increase the yield to 90% under ambient conditions [34–36].

Along with the abovementioned ILs and their applications, ILs like imidazolium-based ones serve as a catalyst in aiding the conversion of CO_2 and help to reduce overpotential during the conversion reaction. Also, these ILs when used during this reaction help to lower the energy of the CO_2^- intermediate which lowers the initial reduction barrier [26,37].

3.3 Biomass conversion to biofuels

As discussed earlier in this chapter, biomass has been a part human activity even before the era of fossil fuels and also the energy produced from biomass is considered to be a very effective and essential alternative. When discussing biomass it mainly includes that obtained from resources like wood, waste, paper energy crops, corn shiver, etc.

The main composition of biomass includes cellulose, hemicellulose, and lignocellulose, making almost 60%–90% of the biomass [2].

ILs with their great solubility properties mainly with organic solvent has increased interest in the mixing of ILs with biomass products/resources to make this alternative greener, i.e., more environment friendly. The conversion of biomass into fuels or fuel additives is mainly the conversion of cellulose, hemicellulose, and lignocellulose into the fuels, and all this requires different kinds of roots and different ILs for their conversion [38].

Fig. 8.6 lists different combinations of ILs which increase the dissolution of cellulose. In 2002, Swatloski et al. [39] found that the addition of ILs resulted in dissolution of cellulose that increased up to almost 25% with microwaves and 10% without microwaves.

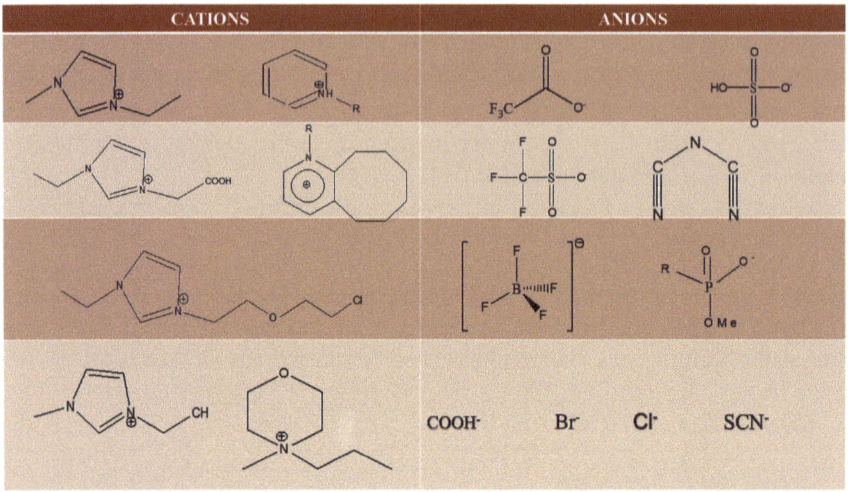

Figure 8.6 Different cations and anions in ionic liquids which can be combined to increase the dissolution of cellulose.

For the dissolution of lignin, which is the third most abundant biopolymer present on Earth, Fu et al. [40] used six kinds of ILs for the conversion process, and among all these, [EMIM][OAc] could dissolve 30% of the lignin IL solution at a temperature of 90°C.

For the dissolution of natural lignocellulose, Argyropoulos et al. [41] showed that use of ILs such as [BMIM][Cl] and [AMIM][Cl] leads to complete dissolution of 8% weight of a dried wood sample at a temperature ranging from 80 to 130°C.

Along with these researchers, many others have also worked on the same concept and considered that [BMIM][Cl], [AMIM][Cl], [EMIM]Cl], and [EMIMI][OAc] are the mostly used ILs, which actively dissolved and took the IL-biomass dissolution process to another level [29,42].

Fig. 8.2 tabulates all the applications of the biomass composition in the different fields mentioned with their conversion of the biomass into its other components and their specific properties and uses of all those components [43].

3.4 Biodiesel

As we look into the field of transportation or an alternate electricity source, we are faced with the production of alternatives to diesel fuels to reduce the pollution caused by the unburned hydrocarbons, sulfur, or particulates from the engine [44]. As an alternative, biodiesel is considered to be the most prominent. For the production of biodiesel, microalgae oil, mahua oil, palm oil, animal oil, cooking oils, etc., can be used, and for commercial biodiesel, the process of transesterification is carried out in the presence of acidic or basic catalysts. For the development of an environment-friendly catalyst, ILs are considered as good media for providing a promising pathway to eco-friendly production.

While producing an environment-friendly diesel, i.e., biodiesel, ILs have been introduced as alternative green reaction media. The favorable properties of ILs include nonvolatility, negligible vapor pressure, and high electrochemical and thermal stability.

3.5 Electrochemical sensors

Over the past few decades, of all the nanomaterials, GNPs and CNTs are considered as to be the most suitable due to their extraordinary physicochemical properties for nonenzymatic glucose. ILs show great applications in many fields because of their unique set of properties. Properties like wide window potential, negligible vapor pressure, high thermal stability, and high conductivity have made them suitable also for sensor applications [31].

GNPs are regarded as cores of composites, CNTs act as support to GNPs, and ILs act as a bridge between GNPs and CNTs connecting one with the other. For this whole composite bucky gel acts as a platform. This explains the relation between GNP, CNT, and IL. As discussed earlier those ILs are composed by the combination of different cations and anions, and the best part about their composition is that according to the specific requirements, we can choose the combination of cations and anions to meet our requirements. Therefore, this study was aimed at the enhancement of the sensor performance for selective and sensitive detection of glucose. For this detection of glucose, different composites of GNPs/CNTs/ILs on a glassy carbon electrode (GPE) were used. For enhancement of the performance of bucky gel, the platform needed for the whole composite needed to be enhanced for sensor enhancement. The modified sensors were constructed by incorporating GNPs to different bucky gels which were made using different ILs with high window potential and thermochemical stability [45]. Fig. 8.7 provides a list of ILs which are being used in electrochemical sensors to enhance their performance.

3.6 Carbon nanotubes

Carbon nanotubes (CNTs) have been assembled and used as an electrode for biosensors as they improve the kinetic transfer to a good extent. The research done by Ajajan and his coworkers concluded that electron oxidation was more rapid on a CNT electrode as compared to all other electrodes. CNTs have some excellent electrochemical properties with specific conductance which make them suitable for areas like sensors, electrochemical actuators, electrocatalysts, and supercapacitors. In the case of CNTs, the potential window availability is limited by the electrolyte because of the inert nature of the carbon. The ability to store the charge, which also determines the performance capabilities, is mainly dependent on the inert electrochemical potential window. Recently ILs have been considered as the most effective electrolytes with a wide electrochemical potential window which makes them useful in the enhancement of the photochemical cells, CNTs, and capacitors with environmental stability.

S.No	Ionic Liquid	Structure
1	1-(4-sulfonylbutyl)-3-methyl imidazolium	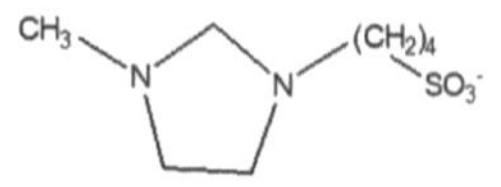
2	1-octyl-3-methyl-imidazolium hexafluorophosphate	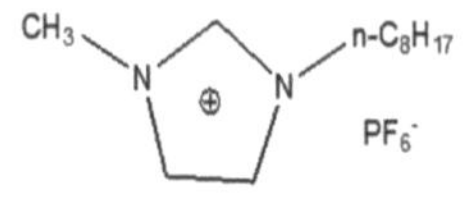
3	Tetrabutyl-ammonium hexafluorophosphate	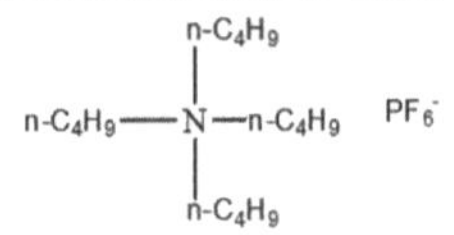
4	N-butyl--methyl-piperdinium bromide	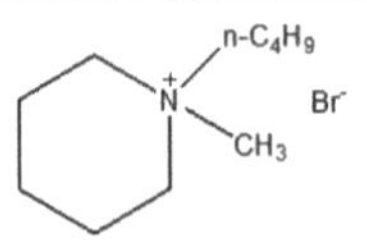
5	N-butyl- -methyl-pyrolidinium hexafluorophospahte	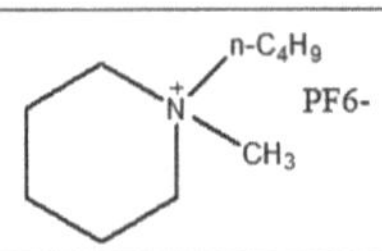

Figure 8.7 Common ionic liquids used in sensors.

3.7 Energy storage devices

The potential application of ILs in electrochemical energy storage and conversion was mainly because of their negligible vapor pressure, thermal stability, transport properties, and interaction with other electrode materials. These properties can be substantially modified to increase the efficiency of the electrochemical process [31]. Also, the use of ILs as a functional electrolyte was the motivation in this field which lead to the development of ILs in the electrochemical energy storage devices as these molten salts have a significant impact on the fabrication of the electrode material, electrolytes, or electrocatalysts.

An initial characterization was carried out for the establishment of the electrochemical window and stability of Ils. Some of the ILs which show great stability and have a high potential window are mentioned in Table 8.2 [46,47].

While considering ILs as electrolytes, the chemical stability of the electrode material also plays an important role. Like in Li−S batteries, the main problem faced is the insulating character of the electroactive material used and also the dissolution of the polysulfide which is formed during the charging and discharging cycles of the battery. Earlier, $LiNO_3$ was used as an additive to overcome this issue, but when ILs have been

Table 8.2 Applications of ionic liquids.

S. No.	Type of ionic liquids	Application
1	Imidazolium-based ILs	In terms of conductivity and Li salt stability, these are considered as an efficient alternative for electrochemical systems
2	Piperidinium-based ILs	These ILs are used as an additive for the electrolytes to enhance their capacitance. They are also used to increase the high-rate performance and life cycle of many cells and increase their electrochemical stability
3	Ammonium-based ILs	Because of their good electrochemical cathodic stability and viscosity, they are used as electrolytes in high-energy electrochemical devices
4	Triazolium-based ILs	Because of their high chemical stability, they are considered as a good electrolyte for energy storage devices

used as an electrolyte in these kinds of batteries, the solubility rate of polysulfide is shown to be very low as compared to the alternative used earlier. Therefore, ILs have been proven to be a good candidate for such types of batteries [48,49].

The large temperature window and the wide electrochemical window of IL offer more environment friendly as well as enable the electrodeposition of reducing metals or semiconductors which could not be deposited from conventional water baths.

3.7.1 Li-ion batteries

The mostly commenly used batteries nowadays are Li-ion batteries, and the electrolyte used for LiB is composed of a mixture of ethylene carbonate, di-methyl carbonate, and $LIPF_6$ salt. The use of this composition has proven to be quite harmful to the environment. Also, many cases of battery explosion and self-destruction of mobile phones and laptops have been noted due to overheating or improper charging of the battery used. Also, the salt used in these kinds of batteries gets hydrolyzed, which forms HF which is also poisonous and shows adverse effects for our environment. Therefore, to overcome this limitation ILs have been studied in efforts to replace the salt, and the new composition fulfills the demands as well as being safe and healthy for the environment. This kind of research on-going to find the best results [50].

3.7.2 Vanadium redox flow batteries

In energy storage technologies, vanadium redox flow batteries (VFBs) are said to be one of the most promising due to their long lifetime, high safety, and efficiency [51].

Over the past few decades, perfluorinated sulfonic acid membranes, such as Nafion from Dupont, have been widely used because of the high proton conductivity and

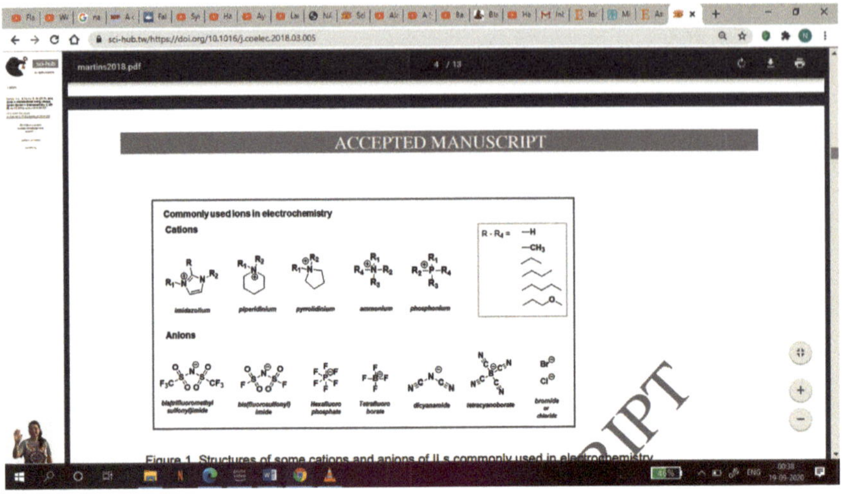

Figure 8.8 Common cations and anions of ionic liquids used in electrochemistry.

chemical stability. Because, in VFBs, ion exchange membrane plays an important role in avoiding the cross-mixing of anolytes and catholytes which completes the circuit by conducting. However, this hindered process used for the construction of these membranes as well as VFBs is high cost and hindered by low selectivity of perfluorinated membranes. To overcome this setback, several approaches have been examined to synthesize anion conductive polymers with a variety of organic cations including quaternary ammonium, imidazolium, and pyridinium as an anion exchange membrane material for these batteries which shows great chemical stability, proton conductivity, and environmental safety [52].

Some of this category of ILs have already been used in energy storage devices and for other purposes in electrochemistry along with their chemical depiction, properties, and applications. Fig. 8.8 and Table 8.2 provide a list of ILs used in electrochemistry along with their applications.

4. Conclusion

While exploring the renewable energy produced from biomass, solar energy, or carbondioxide, many conflicts had been noticed, and to overcome these challenges and to enable the use of green energy easier and less complicated, ILs because of their potential characteristics including solubility toward organic compounds, high thermal stability, negligible vapor pressure, and many more, are suitable for this purpose.

ILs have brought about a revolution in electrochemistry by being used as an electrolyte in many energy storage devices, making them more efficient, with greater potential, environment friendly, as well as being economically friendly. The use of ILs has encouraged a new field in electrochemistry, introducing many more ways to improve the capacitance as well as life cycle of these devices. Among all the electrochemistry applications of ILs, they are set to be a boon for the field of electrochemical substances, providing many more opportunities and paving a way for the fabrication of energy storage devices that are suitable for the environment and enhance their performance.

Based on all the reviews analyzed, we have concluded that ILs offer a great advantage in the dissolution and production of green energy without harming the environment or degrading its resources.

References

[1] Z. Suojiang, S. Jian, Z. Xiaochun, X. Jiayu, M. Qingqing, W. Jianji, Chemical Society and Reviews 43 (2014) 7838–7869.

[2] S.R. Chia, H.C. Ong, K.W. Chew, P.L. Show, S.M. Phang, T.C. Ling, D. Nagarajan, D.J. Lee, J.S. Chang, Sustainable approaches for algae utilisation in bioenergy production, Renewable Energy 129 (2018) 838–852, https://doi.org/10.1016/j.renene.2017.04.001.

[3] M.R. Avhad, J.M. Marchetti, A review on recent advancement in catalytic materials for biodiesel production, Renewable and Sustainable Energy Reviews 50 (2015) 696–718, https://doi.org/10.1016/j.rser.2015.05.038.

[4] K. Ashish, A.S.T. Pramanik, Oriental Journal of Chemistry 29 (2013) 277–283.

[5] J.F. Wishart, Energy applications of ionic liquids, Energy & Environmental Science 2 (9) (2009) 956–961, https://doi.org/10.1039/b906273d.

[6] B. Scrosati, J. Hassoun, Y.K. Sun, Journal of Energy& Environmental Science 4 (2011) 3287–3295.

[7] O. Fontaine, A deeper understanding of the electron transfer is the key to the success of biredox ionic liquids, Energy Storage Materials 21 (2019) 240–245, https://doi.org/10.1016/j.ensm.2019.06.023.

[8] H.M. Srour, N. Giroud, H. Rouault, C.C. Santini, ECS Transactions 41 (2012) 23–25.

[9] A.S. Best, A.I. Bhatt, A.F. Hollenkamp, Journal of the Electrochemical Society 8 (2010).

[10] M. Galinski, L. Andrzej, S. Izabela, Electrochemica acta 26 (2006) 5567–5580.

[11] D.D. Patel, J.-M. Lee, Applications of ionic liquids, The Chemical Record 12 (3) (2012) 329–355, https://doi.org/10.1002/tcr.201100036.

[12] W. Wu, Low-temperature compression-assisted absorption thermal energy storage using ionic liquids, Energy and Built Environment (2020) 139–148, https://doi.org/10.1016/j.enbenv.2019.11.001.

[13] T.R. Cook, D.K. Dogutan, S.Y. Reece, Y. Surendranath, T.S. Teets, D.G. Nocera, Solar energy supply and storage for the legacy and nonlegacy worlds, Chemical Reviews 110 (11) (2010) 6474–6502, https://doi.org/10.1021/cr100246c.

[14] H. Liu, Journal of Materials Science and Technology 35 (2019) 221–234.

[15] F. Sharma, The Journal of Energy and Environmental Science 2 (2012) 951–961.

[16] D.R. Macfarlane, N. Tachikawa, M. Forsyth, J.M. Pringle, P.C. Howlett, G.D. Elliott, J.H. Davis, M. Watanabe, P. Simon, C.A. Angell, Energy applications of ionic liquids, Energy & Environmental Science 7 (1) (2014) 232–250, https://doi.org/10.1039/c3ee42099j.

[17] P. Sterchele, K. Kersten, A. Palzer, J. Hentschel, H.M. Henning, Journal of Applied Energy 258 (2020) 114101, https://doi.org/10.1016/j.apenergy.2019.114101.
[18] M. Watanabe, M.L. Thomas, S. Zhang, K. Ueno, T. Yasuda, K. Dokko, Application of ionic liquids to energy storage and conversion materials and devices, Chemical Reviews 117 (10) (2017) 7190–7239, https://doi.org/10.1021/acs.chemrev.6b00504.
[19] W. Yonggang, S. Yanfang, X. Yongyao, Journal of Chemical Society Reviews 45 (2016) 5925–5950.
[20] K. Harpreet, R.C. Thakur, K. Harsh, K. Arjuna, The Journal of Chemical Thermodynamics 158 (2021).
[21] M. Sandipan, P. Atin, M. Sourindra, Journal of RSC Advances 52 (5) (2015) 41617–41626.
[22] Z.X. Zhang, H.Y. Zhou, L. Yang, K. Tachibana, K. Kamijima, J. Xu, Asymmetrical dicationic ionic liquids based on both imidazolium and aliphatic ammonium as potential electrolyte additives applied to lithium secondary batteries, Electrochimica Acta 53 (14) (2008) 4833–4838, https://doi.org/10.1016/j.electacta.2008.02.008.
[23] Madhulata, S., & Satyen, S. (n.d.). Ionic liquid:New Aspect for Future.
[24] S. Pohlmann, T. Olyschläger, P. Goodrich, J. Alvarez Vicente, J. Jacquemin, A. Balducci, Azepanium-based ionic liquids as green electrolytes for high voltage supercapacitors, Journal of Power Sources 273 (2015) 931–936, https://doi.org/10.1016/j.jpowsour.2014.09.167.
[25] Leta, D., Bhupendra, S. G., & Ming, J. L. (n.d.). Journal of Molecular Liquids. 207, 227–234.
[26] HC, W.H. Ong, A. Chen, Y.Y. Farooq, K.T. Gan, K. Lee, Ashok, Renewable and Sustainable Energy Reviews 113 (2019).
[27] A.R. Hajipour, F. Rafiee, Basic ionic liquids. A short review, Journal of the Iranian Chemical Society 6 (4) (2009) 647–678, https://doi.org/10.1007/BF03246155.
[28] M. Vafaeezadeh, H. Alinezhad, Brønsted acidic ionic liquids: green catalysts for essential organic reactions, Journal of Molecular Liquids 218 (2016) 95–105, https://doi.org/10.1016/j.molliq.2016.02.017.
[29] AM, M. Asim, S. Uroos, M. Naz, G. Sultan, N. Griffin, Muhammad, Journal of Molecular Liquids 287 (2019).
[30] MS, A.S. Islam, A. Ahmed, S. Islam, L.C. Aziz, M. Xian, Mridha, Journal of Chemistry 7 (2014) 3373–3389, https://doi.org/10.1016/j.egyr.2021.05.066.
[31] L. Andreani, J.D. Rocha, Use of ionic liquids in biodiesel production: a review, Brazilian Journal of Chemical Engineering 29 (1) (2012) 1–13, https://doi.org/10.1590/S0104-66322012000100001.
[32] B.H.H. Goh, H.C. Ong, M.Y. Cheah, W.H. Chen, K.L. Yu, T.M.I. Mahlia, Sustainability of direct biodiesel synthesis from microalgae biomass: a critical review, Renewable and Sustainable Energy Reviews 107 (2019) 59–74, https://doi.org/10.1016/j.rser.2019.02.012.
[33] V.M. Ortiz-Martínez, P. Andreo-Martínez, N. García-Martínez, A. Pérez de los Ríos, F.J. Hernández-Fernández, J. Quesada-Medina, Approach to biodiesel production from microalgae under supercritical conditions by the PRISMA method, Fuel Processing Technology 191 (2019) 211–222, https://doi.org/10.1016/j.fuproc.2019.03.031.
[34] T. Ahmad, M. Danish, P. Kale, B. Geremew, S.B. Adeloju, M. Nizami, M. Ayoub, Optimization of process variables for biodiesel production by transesterification of flaxseed oil and produced biodiesel characterizations, Renewable Energy 139 (2019) 1272–1280, https://doi.org/10.1016/j.renene.2019.03.036.

[35] S. Deshmukh, R. Kumar, K. Bala, Microalgae biodiesel: a review on oil extraction, fatty acid composition, properties and effect on engine performance and emissions, Fuel Processing Technology 191 (2019) 232–247, https://doi.org/10.1016/j.fuproc.2019.03.013.
[36] IMR, H.H. Fattah, M.A. Masjuki, M.A. Kalam, H.K. Wakil, M.J. Rashedul, Abedin, Industrial Crops Production 57 (2014) 132–140.
[37] P.K. Devan, N.V. Mahalakshmi, Study of the performance, emission and combustion characteristics of a diesel engine using poon oil-based fuels, Fuel Processing Technology 90 (4) (2009) 513–519, https://doi.org/10.1016/j.fuproc.2009.01.009.
[38] Y. Sun, P. Cooke, H.K. Reddy, T. Muppaneni, J. Wang, Z. Zeng, S. Deng, 1-Butyl-3-methylimidazolium hydrogen sulfate catalyzed in-situ transesterification of Nannochloropsis to fatty acid methyl esters, Energy Conversion and Management 132 (1) (2017) 213–220, https://doi.org/10.1016/j.enconman.2016.10.071.
[39] D.R. MacFarlane, J.M. Pringle, K.M. Johansson, S.A. Forsyth, M. Forsyth, Lewis base ionic liquids, Chemical Communications 18 (2006) 1905–1917, https://doi.org/10.1039/b516961p.
[40] H. Lee, W.S. Shin, J.Y. Jung, C.W. Kim, J.W. Lee, J.H. Kwon, J.W. Yang, Optimization of variables affecting the direct transesterification of wet biomass from Nannochloropsis oceanica using ionic liquid as a co-solvent, Bioprocess and Biosystems Engineering 38 (5) (2015) 981–987, https://doi.org/10.1007/s00449-014-1343-x.
[41] A. Riisagera, R. Fehrmanna, M. Haumannb, P. Wasserscheidb, Supported ionic liquids: versatile reaction and separation media, Topics in Catalysis (2006) 91–102, https://doi.org/10.1007/s11244-006-0111-9.
[42] B. Jiang, Y. Wang, L. Zhang, Y. Sun, H. Yang, B. Wang, N. Yang, Biodiesel production via transesterification of soybean oil catalyzed by superhydrophobic porous poly(ionic liquid) solid base, Energy and Fuels 31 (5) (2017) 5203–5214, https://doi.org/10.1021/acs.energyfuels.7b00443.
[43] H. Zhang, H. Li, H. Pan, A. Wang, S. Souzanchi, C. Xu, Charles), S. Yang, Magnetically recyclable acidic polymeric ionic liquids decorated with hydrophobic regulators as highly efficient and stable catalysts for biodiesel production, Applied Energy 331 (2018) 416–429, https://doi.org/10.1016/j.apenergy.2018.04.061.
[44] J. Wu, Y. Gao, W. Zhang, A. Tang, Y. Tan, Y. Men, B. Tang, New imidazole-type acidic ionic liquid polymer for biodiesel synthesis from vegetable oil, Chemical Engineering and Processing: Process Intensification 93 (2015) 61–65, https://doi.org/10.1016/j.cep.2015.04.005.
[45] J.S. Ye, Y. Wen, W.D. Zhang, L.M. Gan, G.Q. Xu, F.S. Sheu, Nanotechnology 6 (2004).
[46] AS, H.H. Sikitonga, T. Masjuki, H.C. Mahilia, W.T. Ong, M.H. Chong, Boosroh, Renewable and Sustainable Energy Reviews 22 (2013) 346–360.
[47] R.M. Torresi, C.M. Corrêa, T.M. Benedetti, V.L. Martins, Chapter 13: tailoring transport properties aiming for versatile ionic liquids and poly(ionic liquids) for electrochromic and gas capture applications, RSC Smart Materials 2018 (29) (2018) 342–380, https://doi.org/10.1039/9781788010535-00342. Royal Society of Chemistry.
[48] y. Haijun, L. Yuejia, N. Hongli, L. Jianlei, H. Gang, RSC Advances 7 (2017) 33231–33240.
[49] L. Lu, X. Han, J. Li, J. Hua, M. Ouyang, A review on the key issues for lithium-ion battery management in electric vehicles, Journal of Power Sources 226 (2013) 272–288, https://doi.org/10.1016/j.jpowsour.2012.10.060.
[50] M. Park, J. Ryu, W. Wang, J. Cho, Nature Review Materials 2 (2016) 16080, https://doi.org/10.1038/natrevmats.2016.80.

[51] X. Yi, G. Kang, C. Xiaomeng, W. Chenyi, L. Nanwen, L. Lei, Journal of Membrane Science 618 (2021) 118696, https://doi.org/10.1016/j.memsci.2020.118696.
[52] C. Pozo-Gonzalo, M. Kar, E. Jónsson, P.C. Howlett, D.R. MacFarlane, M. Forsyth, Reduction of oxygen in a trialkoxy ammonium-based ionic liquid and the role of water, Electrochimica Acta 196 (2016) 727–734, https://doi.org/10.1016/j.electacta.2016.02.208.

Magnetocaloric materials for green refrigeration

9

D. Swathi[1,2], Nand Kishore Yadav[2], N. Kumar Swamy[2] and N. Pavan Kumar[3]
[1]Department of Physics, Telangana Social Welfare Residential Degree College for Women, Vikarabad, Telangana, India; [2]Department of Physics, ISBM University, Nawapara (Kosmi), Gariyaband, Chhattisgarh, India; [3]Department of Sciences and Humanities, Matrusri Engineering College, Hyderabad, Telangana, India

1. Introduction

In recent times, magnetic refrigeration has been shown as a potential technology in heat pump applications. Previously, vapor compression technology was used in cooling applications, which are based on a mechanism of an expanding and compressing process, of a gas in refrigerators. During this process, harmful gases such as chlorofluorocarbons (CFCs) and hydrofluorocarbons (HFCs) are released and cause damage to the ozone layer. Environmentally friendly cooling technology, i.e., magnetic refrigeration technology, is based on the magnetocaloric effect, which was discovered by Warburg in 1881 [1]. It shows several advantages over conventional gas compression technology (CGC). In addition to cooling, it eliminates the release of CFCs and HFCs and helps minimize the depletion of the ozone layer that impacts the greenhouse effect [2,3]. Magnetic refrigeration shows 20%–50% greater cooling efficiency than CGC. In addition to these advantages, MCE also has biomedical applications such as in the treatment of hyperthermia and drug delivery [4]. Many review articles are available in the literature on MCE properties of materials [5–8]. In this chapter, the focus is made on the biomedical applications of MCE materials and those materials which are suitable for near room-temperature and low-temperature (<100 K) applications. Alloys and oxide materials for room-temperature applications and oxides for low-temperature applications are discussed here in.

2. Magnetocaloric effect

The magnetic refrigeration technology is based on the phenomenon of the magnetocaloric effect (MCE), in which a magnetic material when exposed to a varying magnetic field changes its temperature [9]. In this process, initially the magnetic material which is at temperature T K has random magnetic moments in the absence of an external magnetic field. After exposing it to the magnetic field, the magnetic moments are oriented parallel to the field direction. This causes a decrease in the entropy of the material and an increase in the temperature from T to $T + \Delta T$ (increase in

Green Energy Systems. https://doi.org/10.1016/B978-0-323-95108-1.00008-2

temperature ΔT). Now heat transfer technology can be used to remove the generated heat (ΔT) from the material by passing suitable gases or liquids so that the material temperature again reaches T K under a magnetic field. Afterward, the applied magnetic field is removed under adiabatic conditions. Due to this adiabatic condition, the material uses its internal energy to return to its original random state by decreasing the temperature of the material by ΔT, i.e., from T to $T - \Delta T$. Now, the random orientation of magnetic moments causes an increase in entropy and a decrease in temperature. Passing the fluid at this stage in the refrigerator extracts heat. Performing these four steps periodically we can construct an environmentally friendly magnetic refrigerator. The materials which can exhibit this MCE are called magnetocaloric materials, which include Heusler alloys, perovskite materials, glass ceramics, spinel ferrites, and complex oxides. In addition to the several advantages of MCE, there are certain drawbacks in its usage in real-time applications. Some of these include hysteresis loss and reducing the cooling efficiency due to the aging of the magnetocaloric material during the magnetization/demagnetization cycles. These drawbacks may lead to the discovery of new materials that retain their magnetocaloric properties during magnetization/demagnetization cycles, as shown Fig. 9.1.

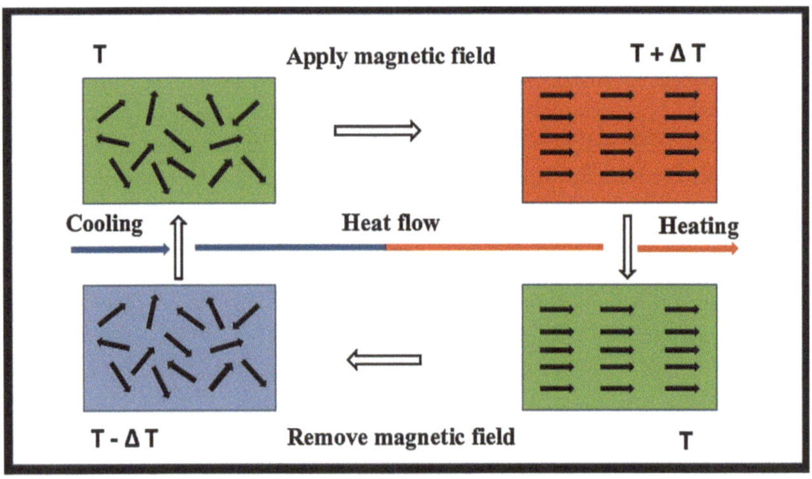

Figure 9.1 Different stages of the magnetocaloric effect.

3. Parameters to calculate the magnetocaloric effect

To calculate the MCE values, we need three parameters, i.e., isothermal entropy change (ΔS), adiabatic temperature change (ΔT), and relative cooling power (RCP). In this process, first, the total entropy can be calculated using the following relation:

$$S(T, H) = \int_0^T \frac{C(T, H)}{T} dT$$

Second, adiabatic temperature change caused by the magnetic field is calculated by the following relation:

$$\Delta T(H,T) = T(S,H) - T(S,0)$$

where *T(S, H)* is the temperature in the field and *T*(*S*, 0) is the temperature at $H = 0$ at constant total entropy.

Finally, the amount of heat transferred between cold and hot sinks in one ideal refrigeration cycle is called the RCP, which represents the cooling efficiency of a given material. It is the product of maximum isothermal entropy change $\Delta S_{\max}$ and the full width at half maximum δT_{FWHM} of the $\Delta S_{\max}$ curve is as follows:

$$\text{RCP} = \Delta S_{\max} \cdot \delta T_{FWHM}$$

Among all the materials, gadolinium (Gd) shows a greater magnetic effect at room temperatures (294 K) [10]. However, it has been limited due to its expensive cost, and so it cannot be used in large-scale applications. But Pecharsky and Gschneidner discovered a giant magnetocaloric effect in $Gd_5Si_2Ge_2$ (Curie temperature 276 K) [11] with $\Delta S_M = -18.5$ J/kg-K and $\Delta T_{ad} = 15$ K at 276 K and $\Delta H = 5$ T in the year 1997.

4. Applications of MCE

Several applications of MCE are cooling technology based on MCE that helps to obtain extremely low temperatures, industrial or commercial refrigeration, heat pump, supermarket refrigeration, large-scale air conditioner, liquor distilling, gas liquefaction, grain drying, sugar refining, chemical processing, etc. However, the majority of the magnetic refrigeration applications are not geared up for commercial use because of its emerging stage of technology. However, the household magnetic refrigerators would likely be very expensive. In addition to this refrigeration kind of application, there is an interesting application of MCE that has been observed in a newer field, i.e., in the biomedical field. The details of a few of these are discussed herein.

Magnetocaloric materials can be used in the treatment of hyperthermia, drug delivery, and in medical implants. When the temperature of a body rises too high and threatens health, it's known as hyperthermia [12]. As a treatment of hyperthermia, some over-the-counter medications can help to reduce the temperature of the body. However, they would be unsuccessful in treating hyperthermia. A change in environment, rehydration, and external cooling efforts (such as cool water or ice packs on the skin) can only reverse hyperthermia. First, conventional magnetic hyperthermia treatments which are being used nowadays have several drawbacks such as a lack of control over the temperature, which may damage the healthy tissue cells in the body and it may also show some side effects such as increased blood pressure. Therefore, controlling temperature is a challenging task here in the treatment of hyperthermia and this can be achieved by using magnetocaloric materials, as they can provide controlled cooling and heating. The benefit of magnetocaloric materials is that as soon as the magnetic field is detached, a cooling effect is produced, unlike in magnetic nanoparticles where overheating persists, even after removal of the magnetic field.

In addition, a high-frequency refrigerator has been described recently using what is called magnetic fluid hypothermia, which consists of heating cancer tumors with particular compatible magnetic nanoparticles delivered directly to the tumor site. To do this, the researchers developed a unique tool to create an alternating high-frequency magnetic field. Recently, Kavita et al. [13] studied human organ-compatible materials which optimized the alloy composition so that the Curie temperature came close to the therapeutic range (i.e., 42−46°C) required for destroying cancer cells [13].

Rejection is one of the problems with artificial implants such as joints, abdominal nets, stents, urinary and biliary ducts, etc. Researchers have proposed applying a special coating to implants consisting of several layers. The first thin-film layer is a negative magnetocaloric material that is cooled in an external magnetic field and then covered with a polymer matrix that absorbs the drug. The polymer matrix is in direct thermal contact with the magnetocaloric material. This entire structure is placed in the body during the operation. When the magnetic field lowers the temperature, the polymer transits in a liquid state and releases the drug at the implantation site. This technique of targeted drug delivery only influences the source of inflammation, leaving the rest of the body unaffected. However, it has to be explored what happens if the coated drug is depleted inside the body. It is also proposed to use magnetocaloric materials as drug-delivery agents during surgery. Komlev et al. [14] projected to use FeRh as a magnetocaloric material for drug delivery. This material is nontoxic like other alloys of iron with platinum group metals. Poly-N-isopropyl-acrylamide (PNIPAm), an implantable Peltier element with a film containing a drug, is supposed to be used as a polymer. Both FeRh and PNIPAm are biocompatible and are widely used in the therapeutic practice of dental prosthetics [14] and the FeRh alloy is a good magnetocaloric material with high magnetic entropy change, which increases the temperature range of the prescribed release of the active constituent [15].

Recently, in a theoretical paper [16], it has been proposed that the magnetic material can exhibit MCE without the application of a magnetic field. It can be achieved in a deformable nonmagnetic matrix having magnetic moments represented by Ising spins embedded into it. This means that the MCE can be controlled solely via the extent of axial deformation of the matrix, whose process affects the interaction between magnetic inclusions. Mechanical metamaterials which exhibit negative Poisson's ratio [16] have to be used to design a formable matrix.

5. Properties of an ideal magnetic refrigerator material

The following properties have to be considered to select the refrigerant material:

(a) High isothermal magnetic entropy change and adiabatic temperature change.
(b) Compounds with high chemical stability and simple preparation methods.
(c) Small lattice entropy and a high density of magnetic entropy for high efficiency.
(d) Low hysteresis loss in magnetic and thermal curves.
(e) Low eddy current loss.
(f) Low heat capacity and high thermal conductivity.
(g) Finally, cost-effective materials.

6. Methods for determination of the magnetocaloric effect

The parameters of MCM are characterized by adiabatic temperature change (ΔT_{ad}) and isothermal magnetic entropy (ΔS_M) change with a varying magnetic field. This can be obtained by two methods, which are direct and indirect methods.

6.1 Direct measurements

The direct method is a straightforward procedure by which we can characterize an MCM, which is isolated adiabatically, calculating the temperature change T_i by placing a temperature sensor on it with a varying magnetic field. After magnetization, the temperature change is T_f. The MCE can be calculated as the difference between the final and initial temperatures ($\Delta T_{ad} = T_f - T_i$). This method was proposed by Weiss in 1921, with a detailed explanation given by Tishin and Spichkin [17]. However, the direct measurements have some disadvantages such as the high uncertainty (7%−15%) [17], low-temperature sensitivity, and measuring in a weak field, etc. The complete characterization of MCE cannot be determined by the direct method as it does not give an isothermal entropy change ΔS.

6.2 Indirect measurements

With indirect methods, the measurements are based on (a) heat capacity under different magnetic fields by varying magnetic fields and (b) M−H curves at different temperatures.

6.2.1 MCE from magnetization measurements

The MCE values in the indirect method can be determined by using basic laws of thermodynamics and Maxwell's equations. The entropy change is given by the fundamental laws of thermodynamics, as follows:

$$\Delta S_M = \int_0^H \left(\frac{\partial S}{\partial H}\right)_T dH \tag{9.1}$$

The relation between the entropy and magnetization is derived from Maxwell's equation, which is as follows:

$$\left(\frac{\partial S(T,H)}{\partial H}\right)_T = \left(\frac{\partial M(T,H)}{\partial T}\right)_H \tag{9.2}$$

In which S is entropy and M is magnetization at temperature T under magnetic field H.

Therefore, the isothermal change in entropy (ΔS), with the magnetic field change from H_1 to H_2, is given by the following relation:

$$\Delta S_M = \int_{H_1}^{H_2} \left(\frac{\partial M(T,H)}{\partial T}\right)_H dH \tag{9.3}$$

To calculate adiabatic change in temperature ΔT_{ad} from the magnetization measurements, knowledge of the heat capacity at a different magnetic field is necessary. ΔT_{ad} with the change in magnetic field can be calculated using the following equation:

$$\Delta T_{ad}(T, \Delta H) = -\int_{H_1}^{H_2} \left(\frac{T}{C(T,H)}\right)_H \left(\frac{\partial M(T,H)}{\partial T}\right)_H dH \tag{9.4}$$

where C is the heat capacity of the system.

6.2.2 MCE from heat capacity measurements

The MCE can be calculated from heat capacity measurements in different magnetic fields. This method is more accurate and reliable than direct methods. In fact, it can provide a complete description of the material which is used for magnetic refrigeration such as heat capacity, entropy, isothermal change in entropy, and adiabatic change in temperature. In this process, the heat capacity is measured in the absence or presence of a magnetic field. From these data, the temperature dependence of entropy is obtained at the corresponding magnetic field. The difference between them will provide a change in entropy ΔS and at constant entropy gives ΔT_{ad}.

7. Magnetocaloric materials for near room-temperature applications

In this section, magnetocaloric materials whose transition temperatures are around room temperature are discussed.

As discussed in Section 9. (Applications), MCE has several applications at room temperature. The well-known magnetocaloric material for room-temperature commercial applications is Gd. The transition temperature of Gd is close to room temperature (294 K) [10]. The isothermal entropy change and adiabatic temperature changes of this rare earth metal are 10 J/kg-K and 11 K at 5 T magnetic field change. Among the various rare earth metals such as Gd [18], Tb [10], Dy [18], Er [19], Tm [20], etc., Gd metal shows a transition at room temperature with high MCE parameters. Researchers have alloyed the Gd metal with various other rare earths, but the MCE values and transition temperatures were not much improved. Susilo et al. [21] studied the Ga-doped Gd alloys and the $(\Delta S_M)_{max}$ values were found to be 7.6 J/kg-K for $Gd_{90}Ga_{10}$ at a 5 T magnetic field. A list of the compounds is given in Table 9.1. Liu et al. [22] studied the $Gd_{50}Co_{50}$ alloy and Fe-added $Gd_{50}Co_{45}$ in which the addition of 5% of Fe

Table 9.1 Magnetocaloric effect of rare earth metals and alloys at room temperature.

S. No.	Sample/ composition	T_c (K)	ΔT (K)	ΔS_{max} (J/kg-K)	References
1	Gd	292	11.2 at 5 T	11.1 at 6 T	[18]
2	Tb	230	10.5 at 6 T	20 at 7.5 T	[10]
3	Dy	180	2.2 t 2 T	14.2 at 6.02 T	[18]
4	Ho	134	4.6 at 6 T	1.8 at 6.02 T	[18]
5	Er	85	3.2 at 6 T	1.2 at 6.02 T	[19]
6	Tm	56	3.1 at 7 T	–	[20]
7	$Gd_{0.8}Ho_{0.2}$	268	10 K at 6.02 T	3.8 at 1 T	[17]
8	Gd_5Si_4	336	8.8 K at 5 T	9 at 5 T	[17]
9	$Gd_5Si_{2.06}Ge_{1.94}$	306	8 K at 5 T	9.5 at 5 T	[17]

decreased the isothermal entropy change from 4.6 to 3.8 J/kg-K and increased the transition temperature to room temperature, i.e., from 268 to 289 K for a magnetic field change of 0–5 T. Even Liu et al. [22] observed a decrease in RCP value from 686 and 673 J/kg [22]. Ao et al. [23] reported the MCE properties of Tb-substituted Gd in which $Gd_{60}Tb_{40}$ showed an isothermal entropy change of 5.15 J/kg-K near room temperature 297 K with a 2 T magnetic field change. The MCE values improved with the substitution of Si to Gd then Tb. Tishin et al. [17] studied the properties of Gd_5Si_4 in which the substitution of Si increased the transition temperature to 336 K and also increased the isothermal entropy change to 9 J/kg-K for a 0–5 T magnetic field change.

Another breakthrough was achieved in the alloy $Gd_5Si_2Ge_2$ in which the Tc is 276 K [11] with a maximum ΔS_{max} of 18.5 J/kg-K and ΔT_{max} of 15 K at 5 T magnetic field change. Also, $Gd_{80}Ge_{15}Si_5$ has shown a ΔS_{max} 11.91 J/kg-K at 260 K with RCP 164 J/kg for a 0–5 T magnetic field change, as studied by Gebara et al. [24]. The list of the compounds discussed is given in Table 9.2.

Table 9.2 Magnetocaloric effect in Gd-based alloys at room temperature.

S. No.	Sample/composition	ΔS_{max} (J/kg-K)	T_c (K)	RCP (J/kg)	H (T)	References
1	$Gd_{90}Ga_{10}$	7.6	285	392	5	[21]
2	$Gd_{50}Co_{50}$	4.6	268	686	5	[22]
3	$Gd_{50}Co_{45}Fe_5$	3.8	289	673	5	[22]
4	$Gd_{60}Tb_{40}$	5.15	297		2	[23]
5	Gd_5Si_4	9	336	–	5	[17]
6	$Gd_5Si_2Ge_2$	18.5	276		5	[11]
7	$Gd_{80}Ge_{15}Si_5$	11.91	260	164	3	[24]
8	$Gd_5Si_2Ge_2B_{0.075}$	18.5	272	351	5	[25]
9	$Gd_5Si_{2.8}Sn_{1.2}$	1.69	301.5	–	1.8	[26]
10	$Gd_5Si_{2.06}Ge_{1.94}$	9.5	306	–	5	[17]
11	$Gd_{80}Ga_{20}$	5.9	286	305	5	[21]

Prabahar et al. [25] studied the MCE properties of boron-added $Gd_5Si_2Ge_2$. It has been observed from their studies that the addition of 5% boron improved the transition temperature and RCP. The maximum isothermal magnetic entropy change obtained in Gd_5Si_2-$Ge_2B_{0.05}$ is 13.6 J/(kg-K) at $\Delta H = 0–5$ T [27], along with a significantly very high RC value of 378 J/kg near room temperature, which makes it more promising for commercial application of magnetic refrigeration technology near room temperature. Zhang et al. [26] studied the properties of $Gd_5Si_{2.8}Sn_{1.2}$, which has an isothermal magnetic entropy change of 1.69 J/kg-K for 1.5 T at a temperature of 301.5 K. Ge added to $Gd_5Si_{2.06}$ improved the isothermal magnetic entropy value and transition temperature, and the values were ΔS_{max} 9.5 J/kg-K at 306 K for a 0–5 T magnetic field change as reported by Tishin et al. [17].

$La(Fe_{0.94}Co_{0.06})_{11.83}Al_{1.17}$ has $\Delta S_{max} = 24$ J/kg-K at a transition temperature of 273 K, as revealed by Hu et al. [28]. Morelli et al. [29] reported that $La_{0.67}Ca_{0.33}MnO_3$ has $\Delta S_{max} = 2.75$ J/kg-K at 260 K under a magnetic field change of 3 T. D. V. Maheshwar Repaka investigated the magnetization and magnetic entropy change in $Pr_{0.58}Sr_{0.42}MnO_3$ [30]. It had a transition temperature of $T_c = 300$ K and magnetic entropy change of 2.33 J/kg-K at 5 T with RCP 65 J/kg. Na-doped $LaMnO_3$ showed a magnetic entropy change of 2.11 J/kg-K with a transition temperature of 342 K with 63 J/kg at 1 T, as reported by Zhang et al. [31]. Zhong [32] investigated the MCE properties of $La_{0.813}K_{0.160}Mn_{0.987}O_3$ compound. It showed a magnetic entropy change of 2.10 J/kg-K and transition temperature of 338 J/kg with RCP 128 J/kg under a 1.5 T magnetic field change. N.T. Hien studied the MCE values of $La_{0.78}Ag_{0.22}MnO_3$ [33]. The transition temperature of 306 K, magnetic entropy change 2.9 J/kg-K, and RCP of 38 J/kg have been observed under a 1 T external magnetic field. Meera [34] reported that the transition temperature of $La_{2/3}Sr_{1/3}MnO_3$ is 370 K, with RCP 41 J/kg, and magnetic entropy change of 1.5 J/kg-K at 1 T. Phan et al. [35] studied the MCE values of $La_{0.7}Ba_{0.3}MnO_3$. The MCE values of this compound are $T_c = 336$ K, $\Delta S_{max} = 1.6$ J/kg-K, and RCP = 36 J/kg at a 1 T magnetic field change. Ayadi reported the MCE properties of $La_{0.6}Sr_{0.2}Ba_{0.2}MnO_3$ [36]. The values are T_c 354 K, ΔS_{max} is 2.26 J/kg-K, with RCP 67 J/kg under 1 T. Phan et al. [37] studied the properties of $La_{0.7}Sr_{0.3}Mn_{0.98}Ni_{0.02}O_3$ which showed a transition temperature of 350 K, magnetic entropy change of 7.65 J/kg-K, with an RCP of 459 J/kg at 7 T.

Recently, a giant MCE of $\Delta S_M = -81.8$ J/kg-K has been reported in $Ni_{50}Mn_{18.5}$-$Cu_{6.5}Ga_{25}$ alloy during reverse martensitic transformation (heating cycle) for a magnetic field change of 9 T at 303 K [38]. The Curie temperature is higher in Mn-rich alloys than Ni-rich alloys of the same series like $Ni_{50}Mn_{25\pm x}Ga_{25-x}$. Zhou et al. [39] studied the MCE properties of a series of $Ni_{50}Mn_{25\pm x}Ga_{25-x}$, in which they observed that the magnetic transition temperature increases as the Mn content increases. The MCE values of $Ni_{50}Mn_{29}Ga_{21}$ are ΔS_M 9.56 J/kg-K 396 K with 8 T. The magnetic and thermodynamic properties of $Ni_{2.19}Mn_{0.81}Ga$ alloy with coupled magnetic and structural phase transitions have been studied by Aliev et al. [40] in which they observed a magnetic entropy change of 4.5 J/kg-K with a Curie temperature of 335 K at 1.5 T. The Curie temperature decreases as Ga is replaced with In. The alloy is $Ni_{59.0}Mn_{23.5}In_{17.5}$, in which the temperature decreases to 246 K and ΔS_{max} decreases to 2.01 J/kg-K with RCP 120 J/kg at a 0–3 T magnetic field change. The maximum value of ΔS_{max} of $Ni_{48}Cu_1Mn_{39}Sb_{12}$ alloy is $\Delta S_{max} = 9.38$ J/kg-K at 291 K for a magnetic field change of 5 T, with a refrigerant capacity of 25.9 J/kg, as reported by Feng et al. [41]. Kainuma et al. [42] reported MCE values of $Ni_{45}Co_5Mn_{36.6}In_{13.4}$ as $\Delta S_{max} = 28.4$ J/kg-K, $T_c = 292$ K at field $\Delta H = 7$ T. Zhukova et al. [43] studied the magnetocaloric properties of $Ni_{59.0}Mn_{23.5}In_{17.5}$ and the estimated magnetocaloric parameters are $\Delta S_{max} = 2.01$ J/kg-K, $T_c = 246$ K under an external magnetic field of $\Delta H = 3$ T with RCP 120 J/kg. Another Mn-based alloy, $MnAs_{0.9}Sb_{0.1}$, has been studied by Wada et al. [44], which exhibited the maximum magnetic entropy change of 35 J/kg-K at the transition temperature (T_c) of 280 K with a change of 5 T magnetic field.

Materials such as $Fe_{88}Zr_8B_4$ exhibit $\Delta S_{max} = 3.3$ J/kg-K, $T_c = 284$ K, and RCP = 646 at field $\Delta H = 5$ T as revealed by Liu et al. [22]. The magnetocaloric properties of $(Fe_{0.8}Mn_{0.2})_3Al$ are $\Delta S_{max} = 0.96$ J/kg-K, $T_c = 350$ K with RCP 400 J/kg at field $\Delta H = 5$ T, as reported by Sharma et al. [45]. The alloy $(Fe_{70}Ni_{30})_{97}Mo_3$ exhibited ΔS_{max} 1.69 J/kg-K, $T_c = 320$ K with RCP 440 J/kg under a magnetic field of 5 T, as reported by Thanveer et al. [46]. Chaudary et al. [47] studied the MCE properties of $Fe_{17}Pr_2$, in which the magnetic entropy change is 4.5 J/kg-K, transition temperature is 292 K with RCP 573 J/kg under magnetic field of 5 T. The alloy $Fe_{17}Y_2$ exhibited ΔS_{max} 4.4 J/kg-K, T_c 301 K with RCP 533 J/kg under a magnetic field of 5 T, as reported by Sanchez Llamazares et al. [48]. The magnetic entropy change of $Fe_{65}Cu_{35}$ is 1.04 J/kg-K, with a transition temperature 360 K and RCP 234 J/kg at a magnetic field change 0–5 T, as reported by Chaudary et al. [47]. The alloy $Fe_{77}Ce_1Si_4Nb_5B_{12}Cu_1$ MCE properties are $\Delta S_{max} = 2.18$ J/kg-K, $T_c = 348$ K with RCP = 271 J/Kg, magnetic field of 5 T, as reported by Tian et al. [49]. The above-described alloys compounds are given in Table 9.3. The Fe-based alloys, $(Fe_{0.72}Cr_{0.28})_3Al$, $Fe_{79}B_{12}Cr_8Gd_1$, exhibit a magnetic entropy change of 1.16, 1.42 J/kg-K and transition temperatures are 330, 355 K with RCP 305, 153 J/kg under an external magnetic field of 5, 1.5 T, as reported by Sharma et al. [45] and Chaudary et al. [51]. Another Fe-based alloy $Fe_{79}Ce_{11}B_6$ has a magnetic entropy change of 3.90 J/kg-K and transition temperature 297 K with RCP 420 J/kg at a 5 T magnetic field change, as shown by Li et al. [52]. The ferrite with the compositional formula $Mn_{0.3}Zn_{0.7}Fe_2O_4$ exhibited ΔS_{max} of 0.88 J/kg-K, RCP of 174 J/kg under a magnetic field of 5 T at a broad temperature range of 30–400 K, as reported by Chaudary et al. [51]. This broad temperature range indicates a high RCP value which is used for magnetic cooling applications, such as self-pumping fluids.

A large magnetic entropy change was found in CMR manganite-based oxides, suggesting that these materials might be exploited for magnetic refrigeration applications. Moreover, an additional advantage with these materials is that they are less expensive as compared with the Gd-based ones. Wang et al. [53] studied the MCE properties of well-known oxide material $La_{0.7}MnO_{3-\delta}$, which has a 1.32 J/kg-K magnetic entropy change at 290 K with RCP of 37 J/kg at 1 T. The effect of Ca doping in $LaMnO_3$ was studied by Vonhelmot et al. [54], who found that Ca-doped $LaMnO_3$ showed an increase in magnetic entropy of 6.4 J/kg-K with increasing doping and the transition temperature decreased to 267 K with a 0–3 T magnetic field change. Another doped material $(La_{0.47}Gd_{0.2})Sr_{0.33}MnO_3$ has a magnetic entropy change of 1.93 J/kg-K, which is lower than that for $La_{2/3}Ca_{1/3}MnO_3$, and has a transition temperature of 289 K at a 2 T field [53]. Phan et al. [8] studied the magnetocaloric properties of the compound $La_{1.6}Ca_{1.4}Mn_2O_7$, and an entropy change of 17 J/kg-K was observed at 270 K under a magnetic field change of 5 T.

Table 9.3 Magnetocaloric effect in various materials at room temperature.

S. No.	Sample/composition	ΔS_{max} (J/kg-K)	T_c (K)	RCP (J/kg)	H (T)	References
1	$La(Fe_{0.94}Co_{0.06})_{11.83}Al_{1.17}$	24	273		2	[28]
2	$La_{0.67}Ca_{0.33}MnO_6$	2.75	260	–	3	[29]
3	$Pr_{0.58}Sr_{0.42}MnO_3$	2.3	300	65	2	[30]
4	$La_{0.835}Na_{0.165}MnO_3$	2.11	342	63	1	[31]
5	$La_{0.813}K_{0.160}Mn_{0.987}O_3$	2.10	338	128	1.5	[32]
6	$La_{0.78}Ag_{0.22}MnO_3$	2.90	306	38	1	[33]
7	$La_{2/3}Ca_{1/3}MnO_3$	6.4	267	–	3	[34]
8	$La_{0.7}Ba_{0.3}MnO_3$	1.6	336	36	1	[35]
9	$La_{0.7}MnO_3$	1.32	290	37	1	[50]
10	$La_{0.6}Sr_{0.2}Ba_{0.2}MnO_3$	2.26	354	67	1	[36]
11	$La_{0.7}Sr_{0.3}Mn_{0.98}Ni_{0.02}O_3$	7.65	350	459	7	[37]
12	$Ni_{50}Mn_{18.5}Cu_{6.5}Ga_{25}$	81.8	303	–	9	[38]
13	$Ni_{2.19}Mn_{0.81}Ga$	4.5	335	–	1.5	[39]
14	$Ni_{50}Mn_{29}Ga_{21}$	9.56	396	–	8	[40]
15	$Ni_{48}Cu_1Mn_{39}Sb_{12}$	9.38	291	25.9	5	[41]
16	$Ni_{45}Co_5Mn_{36.6}In_{13.4}$	28.4	292	–	7	[42]
17	$Ni_{59.0}Mn_{23.5}In_{17.5}$	2.01	246	120	3	[43]
18	$MnAs_{0.9}Sb_{0.1}$	35	280	–	5	[44]
19	$Fe_{88}Zr_8B_4$	3.30	284	646	5	[22]
20	$(Fe_{0.8}Mn_{0.2})_3Al$	0.96	350	400	5	[45]
21	$(Fe_{70}Ni_{30})_{97}Mo_3$	1.69	320	440	5	[46]
22	$Fe_{17}Pr_2$	4.5	292	573	5	[47]
23	$Fe_{17}Y_2$	4.4	301	533	5	[48]
24	$Fe_{65}Cu_{35}$	1.04	360	234	5	[47]
25	$Fe_{77}Ce_1Si_4Nb_5B_{12}Cu_1$	2.39	348	271	5	[49]
26	$(Fe_{0.72}Cr_{0.28})_3Al$	1.16	330	305	5	[45]
27	$Fe_{79}B_{12}Cr_8Gd_1$	1.42	355	153	1.5	[51]
28	$Fe_{79}Ce_{11}B_6$	3.90	297	420.09	5	[52]

8. Materials for low-temperature applications

In addition to room-temperature applications, magnetocaloric materials also have tremendous applications in low temperatures which are essential for cryogenic applications. For this, the materials whose transition temperatures are below <100 K, such as $RMnO_3$ and RMn_2O_5 manganites, the $RTiO_3$ (titanates) and $RFeO_3$ compounds are needed. For $EuTiO_3$, the values of MCE studied by Midya et al. [55] are $\Delta S_{max} = 49$ J/kg-K, under a magnetic field of 7 T at 21 K with RCP of 500 J/kg. Europium doped with La [56] and Ba [57] were reported by Km Rubi and R. Mahendiran et al. They observed that the magnetic entropy changed around 40 J/kg-K at 5 T. Large magnetocaloric properties in single-crystal dysprosium titanate were reported. The values of maximum magnetic entropy change (ΔS_{max}) reach 9.64 and 15.88 J/kg-K for the field changes of 2 and 5 T, respectively. The maximum values of ΔT_{ad} reach 4.14 and 6.79 K for a field change of 2 and 5 T, respectively. For $HoVO_3$, the values of MCE were reported by Balli et al. [58]. The values of $\Delta S_{max} = 17.2$ J/kg-K, under a magnetic field of 7 T at 23.5 K with RCP 800 J/Kg. Y.J. Ke et al. [59] studied the magnetocaloric parameters in $TbFeO_3$ and found that the maximum entropy change is 17.42 J/kg-K, under the magnetic field of 5 T at 9 K. In Tables 9.4 and 9.5, the materials which can exhibit MCE at low temperature are listed.

An MCE study of specific heat measurements in $RMnO_3$ samples was done by Pavan et al. [60]. The typical values of ΔS_{max} at a 5 T magnetic field change are 1.27, 1.08, and 4.0726 J/mol K for $GdMnO_3$, $TbMnO_3$, and $DyMnO_3$, respectively, at below 10 K. The adiabatic temperature difference is obtained from the curves using the specific heat data. For a field change of 5 T, ΔT_{ad} values of $GdMnO_3$, $TbMnO_3$, and $DyMnO_3$ are found to be 2.8, 2.4, and 4.89 K, respectively. Pavan et al. [61] investigated the MCE of Dy-doped $TbMnO_3$ samples. A magnetic field change from 0 to 5 T generated an entropy change of 1 J/mol-K in the vicinity of antiferromagnetic ordering temperature of R^{3+} ions, i.e., in the range of 5–10 K. ΔT_{ad} values of all the samples are found to be around 2 K. MCE properties of a polycrystalline $HoMnO_3$ [62] sample prepared through a sol-gel method have been studied and the ΔS values were 2, 9, and 16 J/kg-K, whereas the ΔT values were 0.6, 3.5, and 8.5 K under 2, 5, and 9 T magnetic fields, respectively. The

Table 9.4 Magnetocaloric effect in La-based ceramic oxides at room temperature.

S. No.	Sample/composition	ΔS_{max} (J/kg-K)	T_c (K)	RCP (J/kg)	H (T)	References
1	$La_{0.7}MnO_{3-\delta}$	1.32	290	37	1	[53]
2	$La_{2/3}Ca_{1/3}MnO_3$	6.4	267	–	3	[54]
3	$(La_{0.47}Gd_{0.2})Sr_{0.33}MnO_3$	1.93	285.9	–	2	[53]
4	$La_{1.6}Ca_{1.4}Mn_2O_7$	17	270	–	5	[8]

Table 9.5 Magnetocaloric effect in alloys at low temperature.

S. No.	Sample/composition	ΔS_{max} (J/kg-K)	T_c (K)	RCP (J/kg)	H (T)	References
1	$EuTiO_3$	49	21	500	7	[55]
2	$HoVO_3$	17.2	23.5	800	7	[58]
3	$GdMnO_3$	4.88	2.80	–	5	[60]
4	$TbMnO_3$	4.12	2.40	–	5	[60]
5	$DyMnO_3$	15.33	4.89	–	5	[60]
6	$Tb_{0.9}Dy_{0.1}MnO_3$	3.43	2.10	37.76	5	[61]
7	$HoMnO_3$	13.1	6.5	320	7	[62]
8	$Gd_{0.9}Ho_{0.1}MnO_3$	2.7	7	–	5	[63]
9	$Tb_{0.9}Ho_{0.1}MnO_3$	1.58	10	–	5	[64]
10	$YMnO_3$	1	42	–	10	[65]

RCP values were found to be 120 and 360 J/kg under 5 and 9 T magnetic fields, respectively. This value is about 80% of the RCP of a single crystalline $HoMnO_3$ sample. In the magnetocaloric studies of $Gd_{1-x}Ho_xMnO_3$ [63] multiferroic compounds, the peak value (ΔS_{max}) increases from 1.25 to 3 J/kg K as Ho doping increases for the case of a 5 T magnetic field condition at below 10 K; this may be attributed to the higher response of Ho^{3+} moments as compared to that of Gd^{3+}. The adiabatic temperature change ΔT values (ΔT_{max}) range from 2.5 to 3.5 K, increasing with the increase in Ho content for a 5 T magnetic field. The shape of ΔS and ΔT curves shows a broad maximum in the vicinity of the antiferromagnetic transition temperature region, which is a characteristic feature of rare earth ion ordering. Also, the peak width appears to increase with increasing magnetic field. This behavior is also attributed to the ordering moments of rare earth ions and their response in different magnetic fields, which is in good correlation with the specific heat studies. The effect of Ho doping on the heat capacity and MCE in $Tb_{1-x}Ho_xMnO_3$ ($x = 0, 0.1,$ and 0.2) has been investigated by Pavan et al. [64]. The typical values ΔS_{max}, which increase with increasing Ho^{3+} concentration are 1.53, 1.6, and 1.8 J/mol K for $x = 0, 0.1,$ and 0.2, respectively, at a 5 T magnetic field at below 10 K. The adiabatic temperature change is obtained from the isentropic curves using the heat capacity data. For a field change of 5 T, ΔT_{ad} values of $TbMnO_3$, $Tb_{0.9}Ho_{0.1}MnO_3$, and $Tb_{0.8}Ho_{0.2}MnO_3$ are found to be around 3–4 K. Swamy et al. [65] studied the MCE properties of $YMnO_3$ compound using specific heat studies, as shown in Fig. 9.2. The curves of isothermal entropy change and adiabatic temperature change showed peaks at around 45 K, the temperature at which $YMnO_3$ became ferrimagnetic. Fig. 9.3 shows the maximum magnitudes of ΔS_M (i.e., ΔS_{Mmax}) and ΔT_{ad} (i.e.ΔT_{ad}^{max}) increased with an increase in the applied magnetic field. ΔS_{Mmax} and

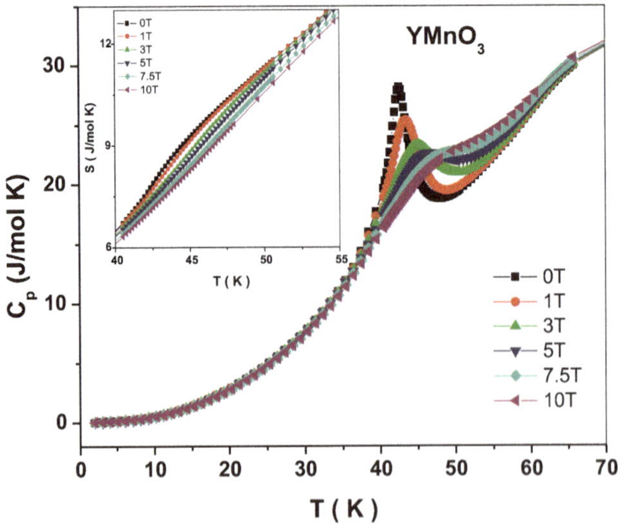

Figure 9.2 Variation of specific heat (Cp) of the $YMnO_3$ sample with temperature (T) under different magnetic fields. Inset shows the calculation of total entropy.

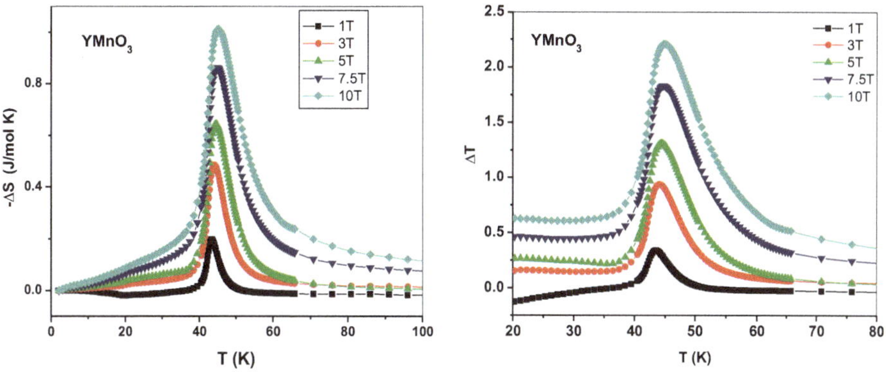

Figure 9.3 Variation of isothermal entropy and adiabatic temperature change with temperature of $YMnO_3$ calculated from heat capacity studies.

ΔT_{ad}^{max} were the highest for a 5 T field and their values were 5.01 J/kg-K (or 1.01 J/mol-K) and 2.2 K, respectively. In a similar way, the magnetic contribution to the entropy (ΔS_M) of $Y_{0.5}Gd_{0.5}MnO_3$ was calculated and the ($-\Delta S_M$) and ΔT_{ad} versus T showed peaks at ~5 K, perhaps due to a Schottky-like anomaly present at that temperature. The ΔS_{Mmax} values were around 0.968 J/mol K and ΔT_{ad}^{max} values were found to be around 1.487 K for the $Y_{0.5}Gd_{0.5}MnO_3$ sample. The total entropy of $Y_{0.3}Gd_{0.7}MnO_3$ [66] has been calculated and the maximum magnetic entropy change ΔS_{Mmax} and ΔT_{ad}^{max} values were found to be ~1.495 J/mol-K and 2 K, respectively, at 10 K.

Wagh et al. [67] studied the properties of $GdMnO_3$(H||c), which has $\Delta S_{max} = 31.8$ J/kg-K, under a magnetic field of 8 T. In the case of a polycrystalline $Tb_{0.9}Sn_{0.1}MnO_3$ sample, Fabris et al. [68] observed a ΔS_{max} value of 4 J/kg-K in a 3 T magnetic field from the magnetization measurements. Magnetocaloric properties of both $TbMnO_3$ and A-site-doped $Tb_{0.67}R_{0.33}MnO_3$ (R = Dy, Y, and Ho) powder samples prepared via a solution route were investigated by Staruch et al. [69]. The calculated values of ΔS_M of these materials were 6.75, 5.25, 5.15, and 6.63 J/kg-K for $TbMnO_3$, $Tb_{0.67}Dy_{0.33}MnO_3$, $Tb_{0.67}Y_{0.33}MnO_3$, and $Tb_{0.67}H_{0.33}MnO_3$, respectively. They also calculated RCP values and the calculated values were 103, 56, 63, and 48 J/kg for $TbMnO_3$, $Tb_{0.67}Dy_{0.33}MnO_3$, $Tb_{0.67}Y_{0.33}MnO_3$, and $Tb_{0.67}H_{0.33}MnO_3$. All the values were calculated at 5 T applied magnetic field change.

Sattibabu et al. [70] doped Er, Ho, Mg, and Sc in $YbMnO_3$ and the MCE properties of those samples were studied. An Mg-doped $YbMnO_3$ sample showed maximum entropy change values of 3.25 and 7.58 J/kg-K at 5 and 10 T magnetic fields, respectively. The adiabatic temperature change values were ~7 K at 5 T and ~16 K at a 10 T applied magnetic field change. The RCP values were 22.98 J/kg at 5 T and 99.57 J/kg at a 10 T field. For the $Yb_{0.9}Sc_{0.1}MnO_3$ sample, the maximum entropy change values were 5.01 and 9.23 J/kg-K at 5 and 10 T fields and the RCP values were 52.43 and 146.28 J/kg at 5 and 10 T fields. The $Yb_{1-x}Ho_xMnO_3$ ($x = 0.1$, 0.2, and 0.3) [71] system shows giant MCE parameters at a high field like 10 T. With increasing Ho concentration from 10% to 30%, a huge increase in RCP from 194 to 328 J/kg-K was observed in this material. Even the small amount of doping of Ho at Yb improves the MCE properties of $YbMnO_3$ effectively. The values of ΔS_M are 10.36, 13.96, and 13.71 for $Yb_{0.9}Ho_{0.1}MnO_3$, $Yb_{0.8}Ho_{0.2}MnO_3$, and $Yb_{0.7}Ho_{0.3}MnO_3$ samples, respectively. The samples with the compositional formula $Yb_{0.9}Er_{0.1}MnO_3$ and $Yb_{0.8}Ho_{0.2}MnO_3$ show around 7.26 J/kg-K isothermal entropy change at 8 T fields. The adiabatic temperature increased from 0.7 K at a 2 T applied magnetic field to 5.4 K at an 8 T field. The RCP values were around 83.81 J/kg-K at an 8 T field.

Balli et al. [58] worked on various multiferroic materials to study their MCE properties. For a single crystalline $TbMn_2O_5$ sample they observed a maximum entropy change of 13.35 J/kg K, for H//a, while it was only 3 J/kg K for H//b, respectively, for 7 T magnetic fields. The entropy change along the hard-axis c was found to be zero. When changing the magnetic field from 0 to 7 T, the RC reached values of 480 J/kg and 47 J/kg along the a and b axes, respectively. The magnetocaloric properties of a single crystal $HoMn_2O_5$ [72] along the easy b axis showed the maximum entropy change values of 13.1 J/kg-K under a field change of 7 T. The RCP values of this sample were 334, 157.5, and 44 J/kg under a magnetic field of 7 T parallel to the b, a, and c axes, respectively. Polycrystalline $HoMn_2O_5$ samples were prepared by Ge et al. [72], who investigated MCE by isothermal magnetization curves. The maximum entropy change at different applied fields and ΔS_M values were 1.7, 6.8, and 7.8 J/kg-K at 2, 5, and 7 T magnetic fields. Moreover, at a 7 T field, this sample showed a large RCP and the value was 216 J/kg.

The MCE values of RMn_2O_5 samples such as $TbMn_2O_5$ have $\Delta S_{max} = 8.94$ J/kg-K, under a magnetic field of 8 T at 4.2 K by Kumar et al. [73]. Balli et al. [58] studied the MCE properties of single crystalline $TbMn_2O_5$ ((H$\|$a) and $HoMn_2O_5$ ((H$\|$b) and showed an entropy change of around 13 J/kg-K under a magnetic field change of 7 T with RCP = 480 J/kg and 334 J/kg, respectively. Ge et al. [72] studied the properties of polycrystalline $HoMn_2O_5$ with $\Delta S_{max} = 7.8$ J/kg-K under a magnetic field of 7 T with RCP = 216 J/kg.

References

[1] G. Wiedemann, Magnetische untersuchungen, Annals of Physics 202 (9) (1865) 1–38, https://doi.org/10.1002/andp.18652020902.

[2] E. Brück, Developments in magnetocaloric refrigeration, Journal of Physics D Applied Physics 38 (23) (2005), https://doi.org/10.1088/0022-3727/38/23/R01.

[3] A. Rowe, Configuration and performance analysis of magnetic refrigerators, International Journal of Refrigeration 34 (1) (2011) 168–177, https://doi.org/10.1016/j.ijrefrig.2010.08.014.

[4] P.W. Egolf, et al., Hyperthermia with rotating magnetic nanowires inducing heat into tumor by fluid friction, Journal of Applied Physics 120 (6) (2016), https://doi.org/10.1063/1.4960406.

[5] V. Franco, J.S. Blázquez, B. Ingale, A. Conde, The magnetocaloric effect and magnetic refrigeration near room temperature: materials and models, Annual Review of Materials Research 42 (2012) 305−342, https://doi.org/10.1146/annurev-matsci-062910-100356.
[6] M. Balli, B. Roberge, P. Fournier, S. Jandl, Review of the magnetocaloric effect in $RMnO_3$ and RMn_2o_5 multiferroic crystals, Crystals 7 (2) (2017), https://doi.org/10.3390/cryst7020044.
[7] N.R. Ram, et al., Review on magnetocaloric effect and materials, Journal of Superconductivity and Novel Magnetism 31 (7) (2018) 1971−1979, https://doi.org/10.1007/s10948-018-4666-z.
[8] M.H. Phan, S.C. Yu, Review of the magnetocaloric effect in manganite materials, Journal of Magnetism and Magnetic Materials 308 (2) (2007) 325−340, https://doi.org/10.1016/j.jmmm.2006.07.025.
[9] A. Gschneidner, V.K. Pecharsky, A.O. Tsokol, Recent developments in magnetocaloric materials, Reports on Progress in Physics 68 (6) (2005) 1479−1539, https://doi.org/10.1088/0034-4885/68/6/R04.
[10] A.M. Tishin, V.K. Pecharsky, K.A. Gschneidner, p3478_1Gd 物性値.pdf, Physical Review B: Condensed Matter 57 (6) (1998) 3478−3490.
[11] K.A. Gschneidner, V.K. Pecharsky, Magnetocaloric materials, Annual Review of Materials Science 30 (1) (2000) 387−429, https://doi.org/10.1146/annurev.matsci.30.1.387.
[12] I.M. Obaidat, V. Narayanaswamy, S. Alaabed, S. Sambasivam, C.V.V. Muralee Gopi, Principles of magnetic hyperthermia: a focus on using multifunctional hybrid magnetic nanoparticles, Magnetochemistry 5 (4) (2019) 67, https://doi.org/10.3390/magnetochemistry5040067.
[13] S. Kavita, et al., On the giant magnetocaloric and mechanical properties of Mn−Fe−P−Si-Ge alloy, Journal of Alloys and Compounds 817 (2020) 153232, https://doi.org/10.1016/j.jallcom.2019.153232.
[14] A.S. Komlev, R.R. Gimaev, V.I. Zverev, Smart magnetocaloric coatings for implants: controlled drug release for targeted delivery, Open Physics 7 (February) (2021) 100063, https://doi.org/10.1016/j.physo.2021.100063.
[15] M.R. Dudek, K.W. Wojciechowski, J.N. Grima, R. Caruana-Gauci, K.K. Dudek, Colossal magnetocaloric effect in magneto-auxetic systems, Smart Materials and Structures 24 (8) (2015) 85027, https://doi.org/10.1088/0964-1726/24/8/085027.
[16] M.R. Dudek, K.K. Dudek, W. Wolak, K.W. Wojciechowski, J.N. Grima, Magnetocaloric materials with ultra-small magnetic nanoparticles working at room temperature, Scientific Reports 9 (1) (2019) 1−10, https://doi.org/10.1038/s41598-019-53617-0.
[17] Y. I, T.A. Spichkin, The_magnetocaloric_effect_and_its_applic, 2003, p. 476.
[18] L. Eyring, K.A. Gschneidner, Handbook on the Physics and Chemistry for Rare Earths, vols. 24, 26, 1999.
[19] J.E. Milton, T.A. Scott, Pressure dependence of the magnetic transitions in dysprosium and erbium, Physics Review 160 (2) (1967) 387−392, https://doi.org/10.1103/PhysRev.160.387.
[20] L.H. Greene, et al., Structural, magnetic and superconducting properties of rare earth/superconductor multilayer films, Superlattices and Microstructures 1 (5) (1985) 407−415, https://doi.org/10.1016/S0749-6036(85)80007-0.
[21] R.A. Susilo, J.M. Cadogan, D.H. Ryan, N.R. Lee-Hone, R. Cobas, S. Muñoz-Pérez, Spin-reorientation in GdGa, Hyperfine Interactions 226 (1−3) (2014) 257−266, https://doi.org/10.1007/s10751-013-0924-4.

[22] G.L. Liu, D.Q. Zhao, H.Y. Bai, W.H. Wang, M.X. Pan, Room temperature table-like magnetocaloric effect in amorphous Gd50Co45Fe5 ribbon, Journal of Physics D Applied Physics 49 (5) (2016) 55004, https://doi.org/10.1088/0022-3727/49/5/055004.
[23] W.Q. Ao, Y.X. Jian, F.S. Liu, X.W. Feng, J.Q. Li, The influence of gallium on magnetocaloric effect in Gd60Tb40 alloys, Journal of Magnetism and Magnetic Materials 307 (1) (2006) 120−123, https://doi.org/10.1016/j.jmmm.2006.03.062.
[24] P. Gebara, M. Hasiak, Determination of phase transition and critical behavior of the as-cast gdgesi-(X) type alloys (where x = ni, nd and pr), Materials 14 (1) (2021) 1−12, https://doi.org/10.3390/ma14010185.
[25] K. Prabahar, N.P. Kumar, D.M. RajKumar, S. Arumugam, M. Manivel Raja, Effect of Boron addition on the microstructure, phase transitions and magnetocaloric properties in Gd-Si-Ge alloy, Intermetallics 96 (February) (2018) 18−24, https://doi.org/10.1016/j.intermet.2018.02.008.
[26] E. Zhang, Y. Chen, T. Zhang, Y. Tang, M. Tu, Magnetocaloric effect of $Gd_5Si_xSn_{4-x}$ alloys, Journal of Magnetism and Magnetic Materials 305 (2) (2006) 410−412, https://doi.org/10.1016/j.jmmm.2006.01.112.
[27] N.P. Kumar, K. Prabahar, D.M.R. Kumar, M.M. Raja, Structural, magnetic and magnetocaloric properties of high-energy ball-milled $Gd_5Si_2Ge_2B_{0.05}$ alloy, Journal of Superconductivity and Novel Magnetism 32 (2) (2019) 319−324, https://doi.org/10.1007/s10948-018-4714-8.
[28] K. Fukamichi, Itinerant-electron metamagnetism theoretical aspects of itinerant-electron metamagnetism, 2001, pp. 5−6.
[29] M. Khlifi, E. Dhahri, E.K. Hlil, Room temperature magnetocaloric effect, critical behavior, and magnetoresistance in Na-deficient manganite $La_{0.8}Na_{0.1}MnO_3$, Journal of Applied Physics 115 (19) (2014) 0−5, https://doi.org/10.1063/1.4879098.
[30] D.V. Maheswar Repaka, M. Aparnadevi, P. Kumar, T.S. Tripathi, R. Mahendiran, Normal and inverse magnetocaloric effects in ferromagnetic $Pr_{0.58}Sr_{0.42}MnO_3$, Journal of Applied Physics 113 (17) (2013) 3−6, https://doi.org/10.1063/1.4793599.
[31] W. Zhong, W. Chen, W. Ding, N. Zhang, Y. Du, Q. Yan, Magnetocaloric properties of Na-substituted perovskite-type manganese oxides, Solid State Communications 106 (1) (1998) 55−58, https://doi.org/10.1016/S0038-1098(97)10239-3.
[32] W. Zhong, et al., Synthesis, structure and magnetic entropy change of polycrystalline $La_{1-x}K_xMnO_{3+\delta}$, Journal of Magnetism and Magnetic Materials 195 (1) (1999) 112−118, https://doi.org/10.1016/S0304-8853(98)01080-4.
[33] N.T. Hien, N.P. Thuy, Preparation and magneto-caloric effect of $La_{1-x}Ag_xMnO_3$ (x = 0.10−0.30) perovskite compounds, Physica B: Condensed Matter 319 (1−4) (2002) 168−173, https://doi.org/10.1016/S0921-4526(02)01118-3.
[34] J. Mira, J. Rivas, L.E. Hueso, F. Rivadulla, M.A. López Quintela, Drop of magnetocaloric effect related to the change from first- to second-order magnetic phase transition in $La_{2/3}(Ca_{1-x}Sr_x)_{1/3}MnO_3$, Journal of Applied Physics 91 (10 I) (2002) 8903−8905, https://doi.org/10.1063/1.1451892.
[35] M.H. Phan, S.C. Yu, N.H. Hur, Y.H. Jeong, Large magnetocaloric effect in a $La_{0.7}Ca_{0.3}MnO_3$ single crystal, Journal of Applied Physics 96 (2) (2004) 1154−1158, https://doi.org/10.1063/1.1762710.
[36] F. Ayadi, et al., Effect of monovalent doping on the physical properties of $La_{0.7}Sr_{0.3}MnO_3$ compound synthesized using sol-gel technique, IOP Conference Series: Materials Science and Engineering 28 (1) (2012), https://doi.org/10.1088/1757-899X/28/1/012054.

[37] M.H. Phan, N.D. Tho, N. Chau, S.C. Yu, M. Kurisu, Large magnetic entropy change above 300 K in a colossal magnetoresistive material $La_{0.7}Sr_{0.3}Mn_{0.98}Ni_{0.02}O_3$, Journal of Applied Physics 97 (10) (2005) 0−5, https://doi.org/10.1063/1.1895472.
[38] S.K. Sarkar, P.D. Babu, A. Biswas, V. Siruguri, M. Krishnan, Krishnan, Giant magnetocaloric effect from reverse martensitic transformation in Ni−Mn−Ga−Cu ferromagnetic shape memory alloys, Journal of Alloys and Compounds 670 (2016) 281−288.
[39] X. Zhou, H. Kunkel, G. Williams, S. Zhang, X. Desheng, Phase transitions and the magnetocaloric effect in Mn rich Ni-Mn-Ga Heusler alloys, Journal of Magnetism and Magnetic Materials 305 (2) (2006) 372−376, https://doi.org/10.1016/j.jmmm.2006.01.029.
[40] A. Aliev, et al., Magnetocaloric effect and magnetization in a Ni-Mn-Ga Heusler alloy in the vicinity of magnetostructural transition, Journal of Magnetism and Magnetic Materials 272−276 (III) (2004) 2040−2042, https://doi.org/10.1016/j.jmmm.2003.12.1363.
[41] W.J. Feng, L. Zuo, Y.B. Li, Y.D. Wang, M. Gao, G.L. Fang, Abnormal e/a-dependence of TM and large inverse magnetocaloric effect in $Ni_{49-x}Cu_xMn_{39}Sb_{12}$ alloys, Materials Science & Engineering B: Solid-State Materials for Advanced Technology 176 (8) (2011) 621−625, https://doi.org/10.1016/j.mseb.2011.02.003.
[42] R. Kainuma, et al., Magnetic-field-induced shape recovery by reverse phase transformation, Nature 439 (7079) (2006) 957−960, https://doi.org/10.1038/nature04493.
[43] V. Zhukova, et al., Magnetic properties and MCE in Heusler-type glass-coated microwires, Journal of Superconductivity and Novel Magnetism 26 (4) (2013) 1415−1419, https://doi.org/10.1007/s10948-012-1978-2.
[44] H. Wada, T. Asano, Effect of heat treatment on giant magnetocaloric properties of $Mn_{1+\delta}As_{1-x}Sb_x$, Journal of Magnetism and Magnetic Materials 290−291 (2005) 703−705, https://doi.org/10.1016/j.jmmm.2004.11.342.
[45] V. Sharma, S. Pattanaik, H. Parmar, R.V. Ramanujan, Magnetocaloric properties and magnetic cooling performance of low-cost $Fe7_{5-x}Cr_xAl_{25}$ alloys, MRS Communications 8 (3) (2018) 988−994, https://doi.org/10.1557/mrc.2018.122.
[46] T. Thanveer, R.V. Ramanujan, S. Thomas, Magnetocaloric effect in amorphous and partially crystallized $Fe_{40}Ni_{38}Mo_4B_{18}$ alloys, AIP Advances 6 (5) (2016), https://doi.org/10.1063/1.4952969.
[47] V. Chaudhary, X. Chen, R.V. Ramanujan, Iron and manganese based magnetocaloric materials for near room temperature thermal management, Progress in Materials Science 100 (2019) 64−98, https://doi.org/10.1016/j.pmatsci.2018.09.005.
[48] J.L. Sánchez Llamazares, C.F. Sánchez-Valdes, P.J. Ibarra-Gaytan, P. Álvarez-Alonso, P. Gorria, J.A. Blanco, Magnetic entropy change and refrigerant capacity of rapidly solidified $TbNi_2$ alloy ribbons, Journal of Applied Physics 113 (17) (2013), https://doi.org/10.1063/1.4794988.
[49] H.C. Tian, X.C. Zhong, Z.W. Liu, Z.G. Zheng, J.X. Min, Achieving table-like magnetocaloric effect and large refrigerant capacity around room temperature in $Fe_{78-x}Ce_x$-$Si_4Nb_5B_{12}Cu_1$ (x = 0−10) composite materials, Materials Letters 138 (2015) 64−66, https://doi.org/10.1016/j.matlet.2014.09.127.
[50] Z. Wang, Q. Xu, J. Sun, J. Pan, H. Zhang, Room temperature magnetocaloric effect of La-deficient bulk perovskite manganite $La_{0.7}MnO_{3-\delta}$, Physica B: Condensed Matter 406 (8) (2011) 1436−1440, https://doi.org/10.1016/j.physb.2011.01.044.
[51] V. Chaudhary, R.V. Ramanujan, Iron oxide-based magnetic nanoparticles for high temperature span magnetocaloric applications, Materials Research Society Symposium Proceedings 1708 (2014) 1−6, https://doi.org/10.1557/opl.2014.527.

[52] Z.B. Li, et al., Tunable Curie temperature around room temperature and magnetocaloric effect in ternary Ce-Fe-B amorphous ribbons, Journal of Physics D Applied Physics 50 (1) (2017) 15002, https://doi.org/10.1088/1361-6463/50/1/015002.

[53] G. Wang, Z.D. Wang, L.D. Zhang, Synthesis and magnetocaloric effect of $(La_{0.47}Gd_{0.2})Sr_{0.33}MnO_3$ polycrystalline nanoparticles, Materials Science & Engineering B: Solid-State Materials for Advanced Technology 116 (2) (2005) 183–188, https://doi.org/10.1016/j.mseb.2004.10.001.

[54] R. Von Helmolt, J. Wecker, T. Lorenz, K. Samwer, Magnetoresistance in $La_{2/3}Ca_{1/3}MnO_{3+\delta}$: dependence on magnetic history, Applied Physics Letters 67 (1995) 2093, https://doi.org/10.1063/1.115089.

[55] A. Midya, et al., Large adiabatic temperature and magnetic entropy changes in $EuTiO_3$, Physical Review B: Condensed Matter 93 (9) (2016) 1–7, https://doi.org/10.1103/PhysRevB.93.094422.

[56] K. Rubi, R. Mahendiran, Magnetocaloric effect in a diluted ferroelectric: $Ba_{0.6}Eu_{0.4}TiO_3$. 2018 IEEE International Magnetic Conference INTERMAG 2018, 2018, pp. 1–4, https://doi.org/10.1109/INTMAG.2018.8508657.

[57] K. Rubi, P. Kumar, D.V. Maheswar Repaka, R. Chen, J.S. Wang, R. Mahendiran, Giant magnetocaloric effect in magnetoelectric $Eu_{1-x}Ba_xTiO_3$, Applied Physics Letters 104 (3) (2014), https://doi.org/10.1063/1.4862981.

[58] M. Balli, B. Roberge, S. Jandl, P. Fournier, T.T.M. Palstra, A.A. Nugroho, Observation of large refrigerant capacity in the $HoVO_3$ vanadate single crystal, Journal of Applied Physics 118 (7) (2015) 1–6, https://doi.org/10.1063/1.4929370.

[59] Y.J. Ke, X.Q. Zhang, Y. Ma, Z.H. Cheng, Anisotropic magnetic entropy change in $RFeO_3$ single crystals(R = Tb, Tm, or Y), Scientific Reports 6 (January) (2016) 1–8, https://doi.org/10.1038/srep19775.

[60] N. Pavan Kumar, G. Lalitha, P. Venugopal Reddy, Specific heat and magnetization studies of $RMnO_3$ (R=Sm, Eu, Gd, Tb and Dy) multiferroics, Physica Scripta 83 (4) (2011), https://doi.org/10.1088/0031-8949/83/04/045701.

[61] N. Pavan Kumar, T. Jitender, V. Laxmi, P. Venugopal Reddy, Thermal, magnetic and electrical properties of $Tb_{1-x}Dy_xMnO_3$ multiferroics, Journal of Magnetism and Magnetic Materials 401 (2016) 860–869, https://doi.org/10.1016/j.jmmm.2015.10.120.

[62] N. Pavan Kumar, et al., Magnetic, thermal and magnetocaloric studies of polycrystalline $HoMnO_3$ compound, Applied Physics A: Materials Science & Processing 125 (7) (2019) 1–7, https://doi.org/10.1007/s00339-019-2776-5.

[63] N.P. Kumar, et al., Magnetocaloric properties of $Gd_{1-x}Ho_xMnO_3$ multiferroic compounds, Journal of Low Temperature Physics 200 (1–2) (2020) 40–50, https://doi.org/10.1007/s10909-020-02471-w.

[64] N.P. Kumar, K. Prabahar, D.M.R. Kumar, M. Palit, S. Arumugam, M.M. Raja, Structural, microstructural and magnetocaloric properties of $Gd_{1-x}Ga_x$ alloys, Materials Research Express 6 (7) (2019), https://doi.org/10.1088/2053-1591/ab1669.

[65] N. Kumar Swamy, et al., Specific heat and magnetocaloric effect studies in multiferroic $YMnO_3$, Journal of Thermal Analysis and Calorimetry 119 (2) (2015) 1191–1198, https://doi.org/10.1007/s10973-014-4223-3.

[66] N. Kumar Swamy, N. Pavan Kumar, M. Gupta, V. Dubey, V. Malik, B.K. Das, Schottky-like anomaly in the low-temperature specific heat of polycrystalline $Y_{0.3}Gd_{0.2}Sr_{0.5}MnO_3$, Materials Physics and Mechanics 18 (1) (2013) 35–41.

[67] A.A. Wagh, K.G. Suresh, P.S. Anil Kumar, S. Elizabeth, Low temperature giant magnetocaloric effect in multiferroic $GdMnO_3$ single crystals, Journal of Physics D Applied Physics 48 (13) (2015) 135001, https://doi.org/10.1088/0022-3727/48/13/135001.

[68] F.W. Fabris, et al., Magnetocaloric effect and magnetic properties of $Tb_{0.9}Sn_{0.1}MnO_3$, Journal of Applied Physics 101 (10) (2007) 0−5, https://doi.org/10.1063/1.2732453.
[69] M. Staruch, L. Kuna, A. McDannald, M. Jain, Magnetic and magnetocaloric properties of $TbMnO_3$ and $Tb_{0.67}R_{0.33}MnO_3$ (R=Dy, Y, and Ho) bulk powders, Journal of Magnetism and Magnetic Materials 377 (2015) 117−120, https://doi.org/10.1016/j.jmmm.2014.10.063.
[70] B. Sattibabu, A.K. Bhatnagar, K. Vinod, A. Mani, Magnetocaloric effect in in doped YbMnO3, Physica B: Condensed Matter 514 (2017) 37−40, https://doi.org/10.1016/j.physb.2017.03.024.
[71] P.K. Naini, J. Satapathy, E.M. Abhinav, A. Srinivas, M.M. Raja, Thermomagnetic properties of Dy0.9Ho0.1MnO3 multiferroics, Physica Status Solidi (A) Applications and Materials Science 217 (17) (2020), https://doi.org/10.1002/pssa.202000138.
[72] H. Ge, X.Q. Zhang, Y.J. Ke, J.L. Jin, Z.X. Liao, Z.H. Cheng, Large reversible magnetocaloric effect in $HoMn_2O_5$, Chinese Physics B 22 (5) (2013), https://doi.org/10.1088/1674-1056/22/5/057502.
[73] N.P. Kumar, S. Bharadwaj, P.V. Reddy, Thermal, magnetic and electric properties of RMn_2O_5 (R=Tb, Dy and Ho) based multiferroics, Multiferroic Materials 1 (1) (2014) 1−8, https://doi.org/10.2478/muma-2014-0001.

A jigsaw puzzle-based reconfiguration technique for enhancing maximum power in partially shaded hybrid photovoltaic array—methodology

10

Palpandian Murugesan [1], Prince Winston David [2], Rajvikram Madurai Elavarasan [3], G.M. Shafiullah [4], Praveen Kumar Balachandran [5] and Thanikanti Sudhakar Babu [6]

[1]Department of Electrical and Electronics Engineering, SCAD College of Engineering and Technology, Cheranmahadevi, Tirunelveli, Tamil Nadu, India; [2]Department of Electrical and Electronics Engineering, Kamaraj College of Engineering and Technology, Virudhunagar, Tamil Nadu, India; [3]Clean and Resilient Energy Systems Laboratory, Texas A&M University, Galveston, TX, United States; [4]Discipline of Engineering and Energy, Murdoch University, Murdoch, Australia; [5]Department of Electrical and Electronics Engineering, Vardhaman College of Engineering, Hyderabad, Telangana, India; [6]Department of Electrical and Electronics Engineering, Chaitanya Bharathi Institute of Technology (CBIT), Hyderabad, Telangana, India

1. Introduction

The gap between energy generation and demand has widened due to the depletion of natural energy sources and rapidly increasing energy demand. India is the second-largest country in terms of population, having a diverse economy and energy consumption profile. For developing countries like India, energy demand is the primary concern for sustainable development. To meet the energy demand, India has doubled its renewable energy installation in the last 5 years [1]. International bodies related to the energy crisis and climate change have insisted that all countries increase their renewable energy capacity. Globally, the contribution of renewable energy generation has increased by 26%, with a major share of 64% in the total electricity generation. Hence, in recent years, the contribution of renewable energy sources has increased compared to coal and natural gas [2]. Among the possible renewable energy sources, the solar energy-based power generation system has been found to be a simple installation, pollution-free, long life, and with less maintenance. Photovoltaics (PV) has become familiar due to the reduction in the price of the module and government subsidies [3].

Green Energy Systems. https://doi.org/10.1016/B978-0-323-95108-1.00009-4

The PV cells are connected in a string to form a module, and the modules are grouped to form an array. The performance of the PV module relies on temperature, irradiation, partial shading, aging, and potential induced degradation, etc. [4,5]. Among these parameters, temperature and irradiation are the dominant factors. Under the Standard Test Condition (STC), the PV array produces a single maximum power point (MPP) on the power—voltage (P—V) characteristics. The conventional maximum power point tracking (MPPT) controller can track this MPP [6].

During the partial shading condition (PSC), the PV array produces multiple peaks on the P—V characteristics due to the shading of trees, chimneys, passing clouds, dust, and bird droppings, etc. [7]. This can be avoided by introducing a bypass diode, which produces multiple peaks on P—V characteristics that can mislead the conventional MPPT controller and soft computing-based MPPT controllers. These are found to be efficient in tracking the global maximum power point (GMPP) [7,8]. During the PSC, mismatch loss (ML) occurs due to the nonlinear characteristics of the PV module and results in a reduction in the efficiency of the PV array [9,10].

The ML can be minimized by adopting an MPPT controller, converter topology, and interconnection schemes. Winston et al. [11,12] and Calcabrini et al. [13] have discussed the PV array interconnection schemes to minimize the ML. Under the STC, series-connected modules in the array produce maximum power. However, under PSCs, the modules are grouped to improve the performance [14]. According to the literature, the conventional interconnection schemes, namely series-parallel (SP), total-cross-tied (TCT), bridge-link (BL), and honey comb (HC) are used to achieve the maximum power and to improve the GMPP. In the SP array, the PV modules are grouped to meet the load requirement. In the TCT array, the ties are connected across the rows of the junction. In the BL array, the modules are connected in a bridge rectifier architecture in series to obtain the required voltage and connected in parallel to obtain the required current. In the HC array, the modules are connected in a hexagonal structure, which combines the advantages of TCT and BL [15—17].

Belhachat et al. [18] conducted a comprehensive study of interconnection schemes, namely series, parallel, SP, TCT, BL, and HC configurations under moving irradiation conditions. The result shows that the TCT configuration has provided superior results compared to other interconnection schemes. In Ref. [19], the authors investigated the interconnection scheme using a genetic algorithm under different PSCs. They concluded that the genetic algorithm provided an improved performance compared with the conventional interconnection schemes. In Ref. [20], the authors compared the SP configuration to the TCT and BL configuration under PSCs. They found that there is no optimum configuration since the performance of the array relies on the shading pattern. From the literature, it is understood that the TCT configuration has yielded superior results under PSCs. The main drawback of the TCT configuration, when the modules are connected in rows, is partial shading that leads to the reduction of output power. In addition, the ties connected across the rows of the junction in the TCT configuration cause more power loss. The BL configuration has higher numbers of series-connected modules and has more power loss (PL) compared to the TCT configuration. The HC configuration has higher numbers of series-connected modules and has more PL compared to the TCT and BL configuration. To overcome this problem, the reconfiguration technique is anticipated.

The reconfiguration technique aims to distribute the effects of partial shading uniformly throughout the array to minimize the ML [21]. In Ref. [22], the authors compared the sudoku-based TCT interconnection with the conventional total-cross-tied (CTCT) configuration. Here, the position of the modules is rearranged based on the sudoku pattern without altering the electrical connection. They reported that the sudoku-based interconnection scheme has yielded improved results. In Refs. [23] and [24], the authors compared the rearranged total-cross-tied (RTCT) configuration with CTCT and found that the RTCT configuration yielded superior results in terms of PL and fill factor (FF). In Ref. [25], the authors proposed a 4 × 4 magic square puzzle pattern to rearrange the module within the array. They concluded that the proposed pattern improved the GMPP under most PSCs. In Ref. [26], the authors proposed a zigzag technique, to rearrange the location of the modules within the PV array based on the proposed technique to reduce PL and improve FF. In Ref. [27], the authors analyzed the mathematical analysis of the HC configuration. In Ref. [28], the authors proposed a two-phase method for rearranging the position of the modules in the PV array and compared it with TCT, partial swarm optimization, and sudoku configurations for improving the power. However, it is necessary to explore the hybrid configuration with a recent and effective reconfiguration technique in a realistic environment. From the literature, it is concluded that the reconfiguration technique improved the GMPP and reduced the local MPP on the P−V characteristics, but these configurations have certain limitations. The effectiveness of the reconfiguration depends on the ability to distribute the effects of partial shading uniformly throughout the array. To overcome this complexity, the hybrid configuration is anticipated.

The hybrid PV array configurations are obtained by combining the advantages of configurations, namely SPTCT, BLTCT, and HCTCT. The authors considered only the shading pattern like progressively horizontal, progressively vertical, and progressively diagonal [25]. In Refs. [25,29−31], the authors analyzed the efficacy of a hybrid bridge-link total-cross-tied (BLTCT) configuration over SP, TCT, BL, and hybrid series-parallel total-cross-tied (SPTCT) configurations. The reduction of ties in the hybrid configuration reduces wiring losses and the complexity of the PV system. From the literature review, it is evident that the hybrid PV array configuration is a better solution for diluting the concentration of the partial shading effects and extracting the maximum power. However, the performance of the hybrid PV array configuration relies on the effectiveness of the reconfiguration technique to disperse throughout the array. Hence, it is necessary to explore the effective reconfiguration technique on hybrid PV array configuration schemes to dilute the concentration of the partial shading effects, thereby extracting more output power from the PV array.

1.1 Novelty of the study

This study presents a reconfiguration technique to rearrange the position of the modules of the 4 × 4 PV array to improve the output power under the PSCs. The contribution of the proposed study is as follows:

- After the investigation, it was found that the jigsaw (JS) puzzle pattern can effectively rearrange the position of the modules within the PV array without altering the electrical connection. Hence, the concentration of the partial shading is diluted by dispersing the shading effects throughout the array.

- To enhance the maximum power of TCT and hybrid configurations such as SPTCT, BLTCT, and honeycomb, total-cross-tied (HCTCT) configurations are rearranged according to the JS puzzle pattern.
- The JS puzzle pattern mitigates the occurrence of local MPP in the P–V characteristics and avoids the need for an MPPT controller to track the GMPP.
- To validate the effectiveness of the JS puzzle-based reconfiguration technique, its performance is compared with the existing reconfiguration techniques like KK [32], SS [32], OE [33], and LS [29] to overcome the following shortcomings.
- In the KK puzzle pattern for a 4 × 4 TCT configuration, most of the modules receive full irradiation (1000 W/m^2) [32].
- In the OE puzzle pattern for a 4 × 4 TCT array, the partially shaded modules in the row or column are rearranged in two rows only; thus, the shading effects are not uniformly distributed throughout the array [33]. In the LS puzzle pattern for a 4 × 4 TCT array, the first column remains unaltered. If the first column is shaded, it results in a reduction of output power. Also, the modules connected in the first row are rearranged in the diagonal, resulting in a reduction of output current [29].

This chapter is organized as follows: Section 10. presents the mathematical modeling of the PV array. Section 10.1 discusses the formation of the JS puzzle and pattern arrangements. Section 10.2 describes the physical relocation of modules in an array according to the JS puzzle pattern. Section 10.3 describes the mathematical analysis of the hybrid configuration. In Section 10.4, a description of the existing puzzle pattern is considered for the investigation, followed by the conclusion in Section 10..

2. Mathematical model of PV array

The single-diode PV model developed in Ref. [34] was designed in a MATLAB Simulink environment. The equivalent circuit of the single-diode PV cell comprises a current source connected in antiparallel with diode, series (R_s), and shunt resistance (R_{sh}).

Apply Kirchhoff's current law for the PV cell model as shown in Fig. 10.1. The PV cell current is given in Eq. (10.1):

$$I = I_{Ph} - I_D - I_{sh} \tag{10.1}$$

where I_{ph} is a photo-current that linearly depends on the incident irradiation, I_D is responsible for the nonlinearity of the PV cell, and I_{sh} is the shunt current of the PV cell. Therefore, the PV cell current is derived as shown in Eq. (10.2):

$$I = \boldsymbol{I_{Ph}} - \boldsymbol{I_o}\left[\exp\left(\frac{\boldsymbol{q(V+IR_s)}}{\boldsymbol{akT}}\right) - 1\right] - \frac{\boldsymbol{V+IR_s}}{\boldsymbol{R_{sh}}} \tag{10.2}$$

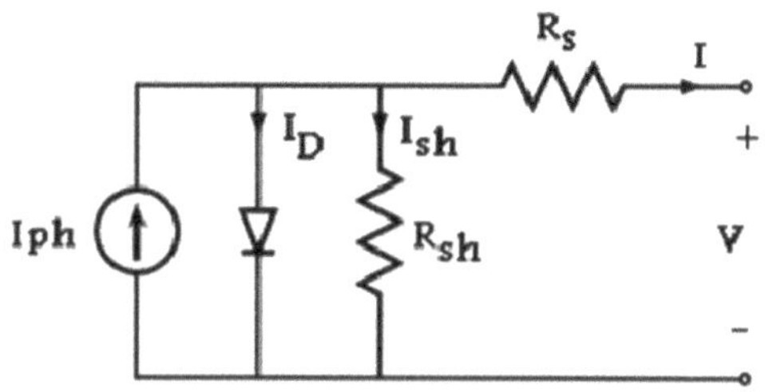

Figure 10.1 Single-diode PV cell.

where I_o is the saturation current of the diode, k is the Boltzmann constant and its value is $1.3806503 \times 10^{-23}$ J/K, q is the electron charge (1.602×10^{-19} C), T is the cell temperature, and a is the ideality constant. The single-diode PV model requires less computation time and fewer unknown parameters compared to the two-diode model of a PV cell.

The PV cells are connected in series to form a PV module and series-connected PV cells are represented as series-connected cells (N_{se}). The module current (I_m) in terms of the module voltage (V_m) with N_{se} cells is connected in series. The module current is represented in Eq. (10.3):

$$\boldsymbol{I_m = I_{ph} - N_p I_o \left[\exp\left(\frac{q(V_m + R_s I_m)}{N_{se} akT}\right) - 1\right] - \frac{V_m + R_s I_m N_{se}}{N_{se} R_{sh}}} \tag{10.3}$$

The PV cells are connected in series to form a module and modules are connected in series and parallel to form a PV array to meet the load requirement. Fig. 10.2 shows the equivalent circuit of the a PV array with series (N_s) and parallel (N_p) connected modules. The PV array current (I_a) in terms of array voltage (V_a) is represented in Eq. (10.4):

$$I_a = N_p I_{ph} - N_p I_o \left[\exp\left(\frac{q\left(V_a + R_s I_a \dfrac{N_s}{N_p}\right)}{N_s akT}\right) - 1\right] - \frac{V_a + R_s I_a \dfrac{N_s}{N_p}}{\dfrac{N_s}{N_p} R_{sh}} \tag{10.4}$$

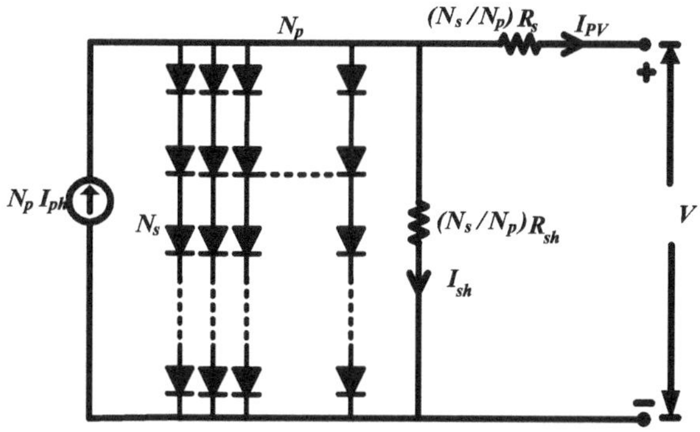

Figure 10.2 Equivalent circuit of a PV array.

Table 10.1 Electrical characteristics of the PV module at STC (1000 W/m^2, 25°C).

Electrical characteristics	Ratings
Maximum power (P_{max})	200W
Open circuit voltage (V_{oc})	32.9V
Voltage at maximum power (V_{mpp})	26.3V
Short circuit current (I_{sc})	8.21A
Current at maximum power (I_{mpp})	7.61A
Temperature coefficient of V_{oc}	-1.23×10^{-1} V/°C
Temperature coefficient of I_{sc}	3.18×10^{-3}A/°C
Number of cells connected in series (N_s)	54
Series resistance (R_s)	0.221 Ω
Parallel resistance (R_{sh})	415.405 Ω
Ideality factor (a)	1.3

The mathematical model of the PV module is obtained with the electrical parameters of the module Kyocera KC200GT, which is represented in Table 10.1. A bypass diode is connected in antiparallel with each module to provide an alternate path under PSC within a series-connected string.

3. Methodology

The JS puzzle belongs to the family of sudoku. The rules of the JS puzzle are similar to standard sudoku. The sudoku puzzle places the numbers from 1 to N in the $m \times n$ grid, without duplicating the same number in the row or column. For example, consider the 4×4 sudoku puzzle pattern, which consists of 16 blocks, constituting four rows and columns. The puzzle consists of four perfect squares, in which each square has numbers from 1 to 4 without duplicating the same number in a row or column.

Similarly, the JS puzzle places the numbers from 1 to N in the $m \times n$ grid, without duplicating the same number in the row or column. The size of the JS puzzle is a perfect square. For example, if the JS puzzle is 4×4, you must use the numbers 1 to 4. The JS puzzle consists of four subarrays with a variety of shapes, which makes it a challenge for the players in solving the puzzle.

3.1 Formation of the proposed jigsaw puzzle arrangement

In this study, a JS puzzle pattern is proposed for rearranging the position of the modules within the 4×4 PV array to mitigate the ML under PSCs. A JS puzzle consists of an $m \times n$ grid, which is similar to an ordinary sudoku puzzle. In the JS puzzle, the squares are irregular in shape, and each of the squares is represented by bold lines. This puzzle places the numbers from 1 to N in the $m \times n$ grid, without duplicating the same number in the row or column. Fig. 10.3A shows the 4×4 JS puzzle pattern,

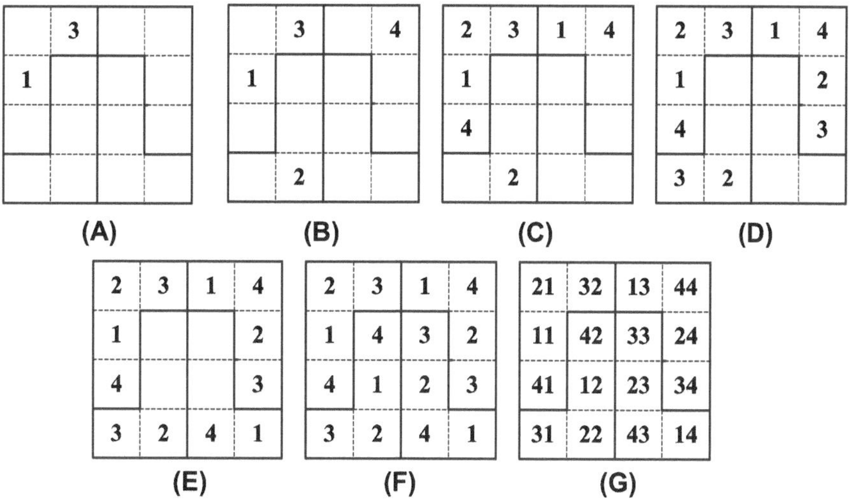

Figure 10.3 (A–G) JS puzzle pattern.

which consists of 16 blocks, which constitutes four rows and columns. The puzzle consists of four irregular squares, in which each square has numbers from 1 to 4 without duplicating the same number in a row or column. The procedure for filling the empty square is as follows:

Step 1: Square 1 has the numbers 1 and 3, which can be filled by 2 and 4 in Fig. 10.3B, and the first row of block 2 is filled with 1 as depicted in Fig. 10.3C.
Step 2: The remaining blocks of square 2 are filled with 2 and 3, and square 3 on the first column with 3 as depicted in Fig. 10.3D.
Step 3: Similarly, square 4 is filled with 1 and 4 as depicted in Fig. 10.3E. Similarly, fill the remaining squares as depicted in Fig. 10.3F. Fig. 10.3G shows the pattern arrangement of the puzzle that has to be considered for further analysis. In the puzzle pattern, the first digit represents the position of the module and the second digit represents the column number. For example in module 21, 2 represents the position of the module, and 1 represents the column number. In the PV array, the position of the modules is rearranged according to the JS puzzle pattern without altering the electrical connection.

3.2 Physical relocation of modules in the PV array

The puzzle arrangement of the proposed JS reconfiguration technique is represented in Fig. 10.3F. This work aims to rearrange the position of the modules within the array according to the JS puzzle pattern. Figs. 10.4 A–D show the CTCT, SPTCT, BLTCT, and HCTCT configurations considered for the analysis. Figs. 10.5A–D show the RTCT, rearranged series-parallel total-cross-tied (RSPTCT), rearranged bridge-link total-cross-tied (RBLTCT), and rearranged honey-comb total-cross-tied (RHCTCT) configurations which are rearranged according to the proposed JS arrangement for dispersing the shading effects uniformly throughout the array.

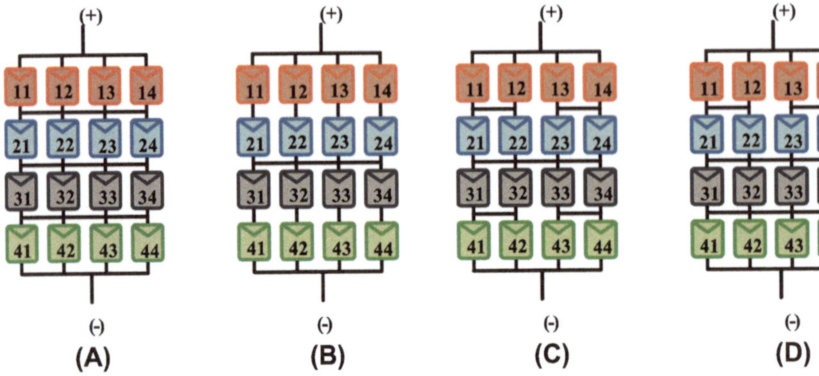

Figure 10.4 PV configuration: (A) CTCT; (B) SPTCT; (C) BLTCT; (D) HCTCT.

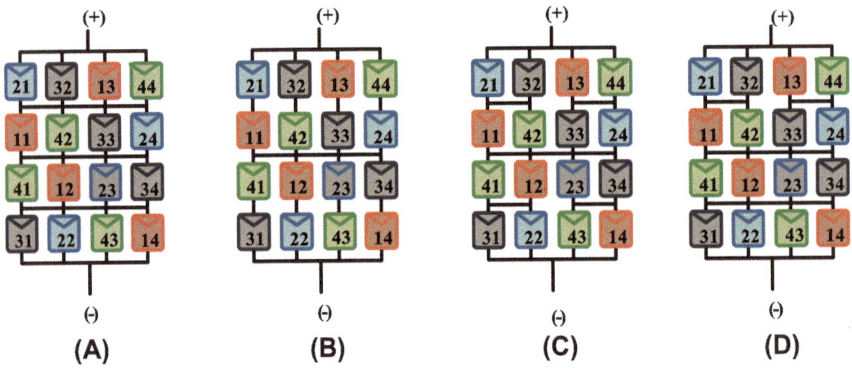

Figure 10.5 JS puzzle pattern for: (A) RTCT; (B) RSPTCT; (C) RBLTCT; (D) RHCTCT.

3.3 Mathematical analysis of the PV array configurations

The mathematical analysis of the TCT and hybrid configurations is derived for better understanding. The expressions for the voltage, current, and power of the TCT and hybrid configuration are derived as follows.

3.3.1 TCT PV array configuration

The TCT configuration is derived from the SP configuration by connecting ties across the rows of the junction. The array current is the sum of the current produced by the modules connected in a row. The row voltage is equal to the open-circuit voltage of the single module. The array voltage is the sum of the voltage across all rows [35]. The current flowing through each module and voltage across each module in the TCT configuration are represented in Fig. 10.6. The array voltage, current, and power of the TCT configuration are given by the following equations:

$$I_{array} = iI_{string} = 4I_m \tag{10.5}$$

$$V_{array} = jV_{string} = 4V_m \tag{10.6}$$

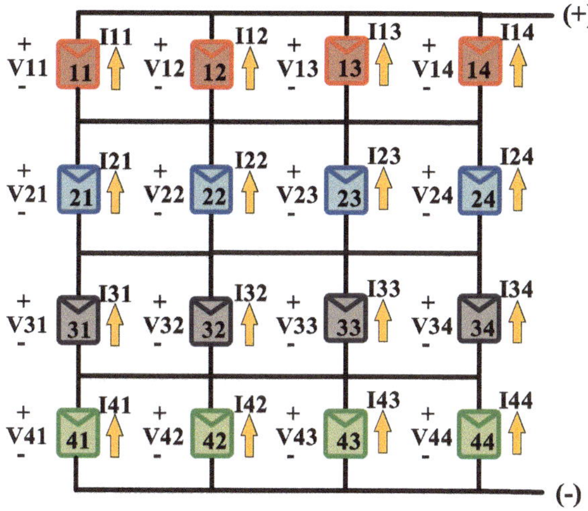

Figure 10.6 Mathematical analysis of the TCT configuration.

$$P_{\text{array}} = (V_{\text{array}})(I_{\text{array}}) = 16V_m I_m \tag{10.7}$$

where i is the number of modules connected in parallel in a string and j is the number of strings connected in series.

3.3.2 SPTCT PV array configuration

The SPTCT configuration is the combination of the SP and TCT configurations. In the SPTCT configuration, the rows are cross-linked at every alternate row. This has a lower number of interconnections compared to BLTCT and HCTCT. However, SPTCT generates less power than hybrid configurations [36]. The current flowing through each module and voltage across each module in the SPTCT configuration are represented in Fig. 10.7. The array voltage, current, and power of the SPTCT configuration are given by the following equations:

$$I_{\text{array}} = I_{11} + I_{12} + I_{13} + I_{14} = 4I_m \tag{10.8}$$

$$V_{\text{array}} = \sum_{i}^{4} V_i = 4V_m \tag{10.9}$$

$$P_{\text{array}} = (V_{\text{array}})(I_{\text{array}}) = 16V_m I_m \tag{10.10}$$

where i is the row of the PV array.

3.3.3 BLTCT PV array configuration

The BLTCT configuration is the combination of the BL and TCT configurations. This configuration is an alternative to the TCT configuration, which can be obtained by

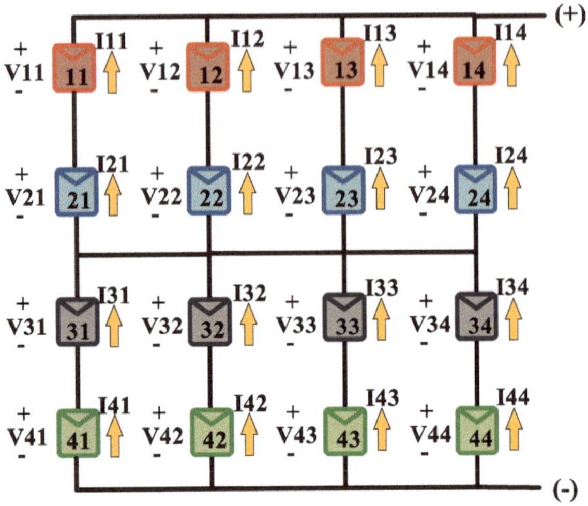

Figure 10.7 Mathematical analysis of the SPTCT configuration.

reducing the number of ties connected among the modules. Hence, the stress on the PV module is minimized [37]. The current flowing through each module and the voltage across each module in the BLTCT configuration are represented in Fig. 10.8. The array voltage, current, and power of the BLTCT configuration are given by the following equations.

$$I_{\text{array}} = I_{11} + I_{12} + I_{13} + I_{14} = 4I_m \tag{10.11}$$

$$V_{\text{array}} = \sum_{i}^{4} V_i = 4V_m \tag{10.12}$$

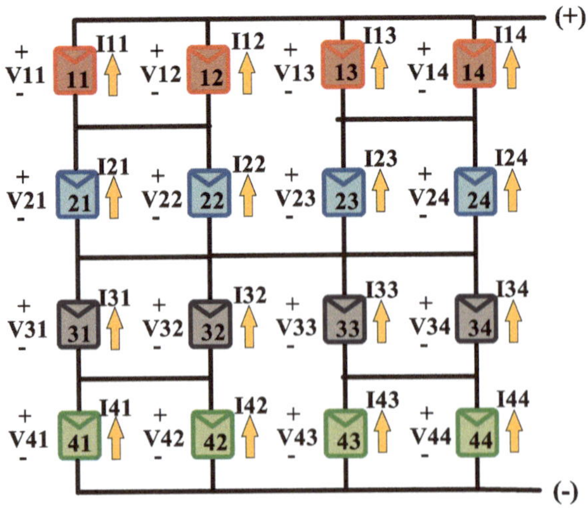

Figure 10.8 Mathematical analysis of the BLTCT configuration.

$$P_{\text{array}} = \left(V_{\text{array}}\right)\left(I_{\text{array}}\right) = 16V_m I_m \tag{10.13}$$

where i is the row of the PV array.

3.3.4 HCTCT PV array configuration

The HCTCT configuration is the combination of the HC and TCT configurations. In the HC configuration, the first and second rows are connected in the HC structure and the third and fourth rows are connected in the TCT configuration. The number of parallel-connected modules in a row is high compared to the BLTCT [27]. The current flowing through each module and the voltage across each module in the HCTCT configuration are represented in Fig. 10.9. The array voltage, current, and power of the HCTCT configuration are given by the following equations:

$$I_{\text{array}} = I_{11} + I_{12} + I_{13} + I_{14} = 4I_m \tag{10.14}$$

$$V_{\text{array}} = \sum_{i}^{4} V_i = 4V_m \tag{10.15}$$

$$P_{\text{array}} = \left(V_{\text{array}}\right)\left(I_{\text{array}}\right) = 16V_m I_m \tag{10.16}$$

where i is the row of the PV array.

3.4 Description of the existing puzzle patterns

The conventional 4 × 4 PV array is rearranged according to the proposed JS puzzle pattern as depicted in Figs. 10.4A–D and Figs. 10.5A–D. To investigate the effectiveness of the proposed puzzle pattern, a comparison is made with the RTCT, RSPTCT,

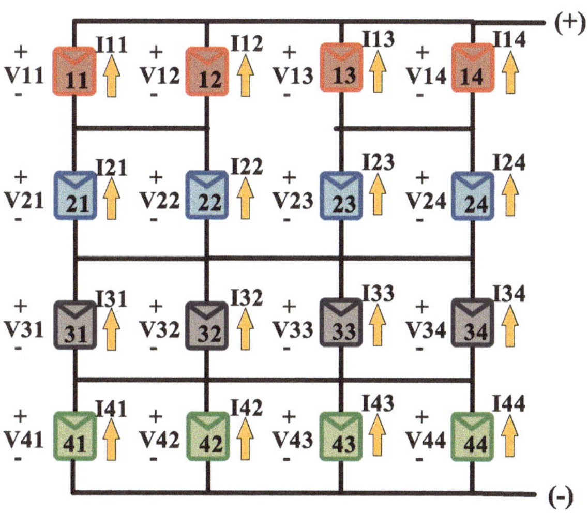

Figure 10.9 Mathematical analysis of the HCTCT configuration.

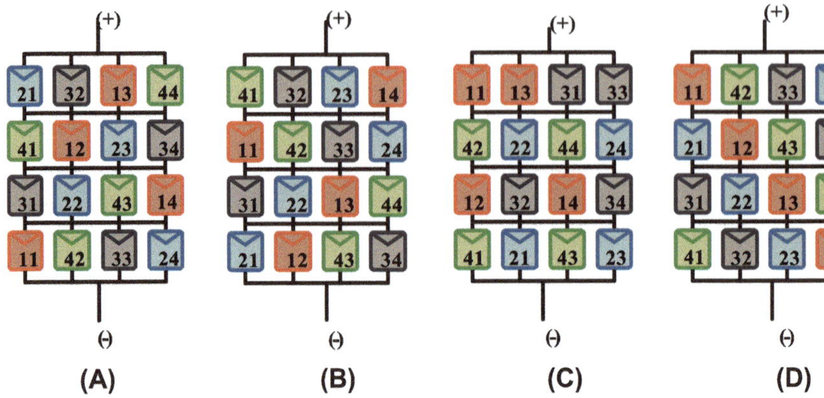

Figure 10.10 Existing puzzle pattern: (A) KK; (B) SS; (C) OE; (D) LS.

RBLTCT, and RHCTCT configurations. The I−V and P−V characteristics of these rearranged configurations modeled in the MATLAB/Simulink environment are utilized for comparing the performance of the proposed puzzle pattern under PSCs. A comparison has also been made with the existing reconfiguration techniques, namely KK, SS, OE, and LS, in mitigating the ML under PSCs. All these configurations are illustrated in Figs. 10.10A−D.

Under uniform irradiation conditions, the entire 4 × 4 PV array configuration produces a maximum power of 3200 W. During the PSCs, the output power generated by the PV array decreases due to the ML. To mitigate the ML, the PV array is rearranged with a hybrid combination. This hybrid interconnection yields superior results to the CTCT interconnection during PSCs. The JS puzzle pattern is applied to the hybrid interconnection to rearrange the physical location of the module within the array. Hence, the concentration of the shading is reduced, thereby distributing the shading effects throughout the array.

4. Conclusion

In this chapter, a JS puzzle pattern-based reconfiguration technique is proposed to extract the maximum power on the hybrid PV array configuration under PSCs. The mathematical model of the PV array is derived. The procedure for the formation of the 4 × 4 JS puzzle pattern is clearly described. The physical location of the modules in the hybrid PV array configuration is rearranged according to the JS puzzle pattern. The mathematical analysis of the PV array configuration is derived for a better understanding. To validate the performance of the proposed JS puzzle pattern, its performance is compared with the existing puzzle pattern, like KK, SS, OE, and LS. The partial shading pattern considered for the analysis and results of the PV array configuration is elaborated in the next chapter.

References

[1] World Energy Council: World Energy Trilemma Index, Available online: https://www.worldenergy.org/wp-content/uploads/2018/10/World-Energy-Trilemma Index-2018.pdf, 2018. (Accessed 21 March 2021).

[2] REN21, Renewables, Global Status Report 2019, 2019. Available online: https://www.ren21.net/wpcontent/uploads/2019/05/gsr_2019_full_report_en.pdf. (Accessed 21 March 2021).

[3] P. Laha, B. Chakraborty, P.A. Østergaard, Electricity system scenario development of India with import independence in 2030, Renewable Energy 151 (2020) 627−639.

[4] L. Tang, X. Wang, W. Xu, C. Mu, B. Zhao, Maximum power point tracking strategy for photovoltaic system based on fuzzy information diffusion under partial shading conditions, Solar Energy 220 (2021) 523−534.

[5] S. Cheng, T.-Y. Qin, L.-B. Jing, Multi-objective optimization of accommodating distributed generation considering power loss, power quality, and system stability, Journal of Industrial and Production Engineering 31 (3) (2014) 146−154.

[6] M.N. Bhukya, V.R. Kota, A quick and effective MPPT scheme for solar power generation during dynamic weather and partial shaded conditions, Engineering Science and Technology 22 (3) (2019) 869−884.

[7] A.S. Yadav, V. Mukherjee, Conventional and advanced PV array configurations to extract maximum power under partial shading conditions: a review, Renewable Energy 178 (2021) 977−1005.

[8] K. Bavarinos, A. Dounis, P. Kofinas, Maximum power point tracking based on reinforcement learning using evolutionary optimization algorithms, Energies 14 (2021) 335.

[9] G. Kampitsis, E. Batzelis, R.V. Erp, E. Matioli, Parallel PV configuration with magnetic-free switched capacitor module-level converters for partial shading conditions, Energies 14 (2021) 456.

[10] S. Haq, S.P. Biswas, M.K. Hosain, M.A. Rahman, M.R. Islam, S. Jahan, A modular multilevel converter with an advanced PWM control technique for grid-tied photovoltaic system, Energies 14 (2021) 331.

[11] D. Prince Winston, S. Kumaravel, B. Praveen Kumar, S. Devakirubakaran, Performance improvement of solar PV array topologies during various partial shading conditions, Solar Energy 196 (2020) 228−242.

[12] D. Prince Winston, G. Karthikeyan, M. Pravin, O. JebaSingh, A.G. Akash, S. Nithish, S. Kabilan, Parallel power extraction technique for maximizing the output of solar PV array, Solar Energy 213 (2021) 102−117.

[13] A. Calcabrini, R. Weegink, P. Manganiello, M. Zeman, O. Isabella, Simulation study of the electrical yield of various PV module topologies in partially shaded urban scenarios, Solar Energy 225 (2021) 726−733.

[14] S.R. Pendem, S. Mikkili, Modelling and performance assessment of PV array topologies under partial shading conditions to mitigate the mismatching power losses, Solar Energy 160 (2018) 303−321.

[15] S.N. Nnamchi, C.O.C. Oko, F.L. Kamen, O.D. Sanya, Mathematical analysis of interconnected photovoltaic arrays under different shading conditions, Cogent Eng 5 (2018) 1507442.

[16] S. Mohammadnejad, A. Khalafi, S. Morteza Ahmadi, Mathematical analysis of total-cross-tied photovoltaic array under partial shading condition and its comparison with other configurations, Solar Energy 133 (2016) 501−511.

[17] J.C. Teo, R.H. Tan, V.K. Ramachandaramurthy, C. Tan, Impact of partial shading on the P–V characteristics and the maximum power of a photovoltaic string, Energies 11 (2018) 1860.

[18] F. Belhachat, C. Larbes, Modeling analysis and comparison of solar photovoltaic array configurations under partial shading conditions, Solar Energy 120 (2015) 399–418.

[19] S.N. Deshkar, S.B. Dhale, J.S. Mukherjee, T.S. Babu, N. Rajasekar, Solar PV array reconfiguration under partial shading conditions for maximum power extraction using genetic algorithm, Renewable and Sustainable Energy Reviews 43 (2015) 102–110.

[20] H. Braun, S.T. Buddhaa, V. Krishnan, Topology reconfiguration for optimization of photovoltaic array output, Sustain Energy Grids Networks 6 (2016) 58–69.

[21] T.S. Babu, J.P. Ram, T. Dragičević, M. Miyatake, F. Blaabjerg, N. Rajasekar, Particle swarm optimization based solar PV array reconfiguration of the maximum power extraction under partial shading conditions, IEEE Transactions on Sustainable Energy 9 (1) (2018) 74–85.

[22] B. InduRani, G.S. Ilango, C. Nagamani, Enhanced power generation from PV array under partial shading conditions by shade dispersion using Su Do Ku configuration, IEEE Transactions on Sustainable Energy 4 (3) (2013) 594–601.

[23] S. Vijayalekshmy, G.R. Bindu, S.R. Iyer, Performance improvement of partially shaded photovoltaic arrays under moving shadow conditions through shade dispersion, Journal of The Institution of Engineers (India) Series B 96 (2015) 1–7.

[24] S.R. Potnuru, D. Pattabiraman, S.I. Ganesan, N. Chilakapati, Positioning of PV panels for reduction in line losses and mismatch losses in PV array, Renewable Energy 78 (2015) 264–275.

[25] A.S. Yadav, R.K. Pachauri, Y.K. Chauhan, S. Choudhury, R. Singh, Performance enhancement of partially shaded PV array using novel shade dispersion effect on magic-square puzzle configuration, Solar Energy 144 (2017) 780–797.

[26] S. Vijayalekshmy, G.R. Bindu, S. Rama Iyer, A novel Zig-Zag scheme for power enhancement of partially shaded solar arrays, Solar Energy 135 (2016) 92–102.

[27] M. Premkumar, U. Subramaniam, T.S. Babu, R.M. Elavarasan, L. Mihet-Popa, Evaluation of mathematical model to characterize the performance of conventional and hybrid PV array topologies under static and dynamic shading patterns, Energies 13 (2020) 3216.

[28] S.P. Dhanup, N. Rajasekar, J.P. Ram, V.K. Chinnaiyan, Design and testing of two phase array reconfiguration procedure for maximizing power in solar PV systems under partial shade conditions (PSC), Energy Conversion and Management 178 (2018) 92–110.

[29] R.K. Pachauri, A.S. Yadav, Y.K. Chauhan, A. Sharma, V. Kumar, Shade dispersion-based photovoltaic array configurations for performance enhancement under partial shading conditions, Int. Trans. Electr. Energy Syst. 28 (2018) 2556–2587.

[30] H. Renaudineau, A. Houari, J.P. Martin, S. Pierfederici, T.F. Meibody, B. Gerardin, A new approach in tracking maximum power under partially shaded conditions with consideration of converter losses, Solar Energy 85 (2011) 2580–2588.

[31] A.S. Yadav, R.K. Pachauri, Y.K. Chauhan, Comprehensive Investigation of PV arrays with puzzle shade dispersion for improved performance, Solar Energy 129 (2016) 256–285.

[32] G. Sai Krishna, T. Moger, Enhancement of maximum power output through reconfiguration techniques under non-uniform irradiance conditions, Energy 187 (2019) 115917.

[33] I. Nasiruddin, S. Khatoon, M.F. Jalil, R.C. Bansal, Shade diffusion of partial shaded PV array by using odd-even structure, Solar Energy 181 (2019) 519–529.

[34] M.G. Villalva, J.R. Gazoli, E.R. Filho, Comprehensive approach to modeling and simulation of photovoltaic arrays, IEEE Transactions on Power Electronics 24 (5) (2009) 1198–1208.

[35] R.P. Vengatesh, S.E. Edward Rajan, Analysis of PV module connected in different configurations under uniform and non-uniform solar radiations, International Journal of Green Energy 13 (14) (2016) 1507–1516.
[36] M. Dhimish, V. Holmes, B. Mehrdadi, M. Dales, B. Chong, L. Zhang, Seven indicators variations for multiple PV array configurations under partial shading and faulty PV conditions, Renewable Energy 113 (2017) 438–460.
[37] B. Nayak, A. Mohapatra, P. Das, Optimal hybrid array configuration scheme to reduce mismatch losses of photovoltaic system, in: Proceedings of the Second International Conference on Electrical, Computer and Communication Technologies (ICECCT), 2017, pp. 1–7.

A jigsaw puzzle-based reconfiguration technique for enhancing maximum power in partially shaded hybrid photovoltaic array—Implementation

11

Palpandian Murugesan [1], *Prince Winston David* [2], *Rajvikram Madurai Elavarasan* [3], *G.M. Shafiullah* [4] *and Praveen Kumar Balachandran* [5]
[1]Department of Electrical and Electronics Engineering, SCAD College of Engineering and Technology, Cheranmahadevi, Tirunelveli, Tamil Nadu, India; [2]Department of Electrical and Electronics Engineering, Kamaraj College of Engineering and Technology, Virudhunagar, Tamil Nadu, India; [3]Clean and Resilient Energy Systems Laboratory, Texas A&M University, Galveston, TX, United States; [4]Discipline of Engineering and Energy, Murdoch University, Murdoch, Australia; [5]Department of Electrical and Electronics Engineering, Vardhaman College of Engineering, Hyderabad, Telangana, India

1. Introduction

The PV modules are grouped together to form a PV array to meet the load requirements. During the uniform irradiation condition, the PV array produces a uniform row current, whereas partial shading may result in uneven row current results in multiple peaks in P−V characteristics [1]. Based on the literature review discussed in the previous chapter, the hybrid configuration is found to be optimal under partial shading conditions (PSCs). The performance of the PV configuration lies on the reconfiguration technique [2−5]. To validate the effectiveness of the proposed jigsaw (JS) puzzle pattern, its performance is compared with the existing puzzle patterns like ken-ken (KK) [6], skyscraper (SS) [6], odd-even (OE) [7], and Latin square (LS) [8]. The most commonly occurring shading patterns like short narrow, short wide, long narrow, and long wide are considered. The performance of the various configurations is analyzed in terms of global maximum power point (GMPP), power loss (PL), mismatch loss (ML), fill factor (FF), execution ratio (ER), and performance enhancement ratio (PE). The effectiveness of the JS puzzle arrangements is validated by experimental verification.

This chapter is organized as follows. Section 11. presents the partial shading scenario. Section 11. describes the performance parameter considered for the analysis.

Green Energy Systems. https://doi.org/10.1016/B978-0-323-95108-1.00011-2

Section 11. describes the results and discussion followed by experimental verification in Section 11. The conclusion is given in Section 11.

2. Partial shading scenario

In this study, the effectiveness of the JS puzzle pattern is verified by comparing the performance of the different conventional and hybrid configurations under PSCs. Four types of shading conditions, namely short narrow, short wide, long narrow, and long wide are considered. Under the PSCs, it is assumed that the shaded module receives irradiation of 100, 200, 300, 400, and 500 W/m^2, and the unshaded module receives irradiation of 1000 W/m^2.

This type of shading condition is applied to the 4 $\times$ 4 JS pattern. The partial shading dispersion effect on the JS puzzle pattern is applied under different configurations. Similarly, the partial shading dispersion effect is applied to the existing KK, SS, OE, and LS patterns. The performances of the existing and JS puzzle pattern are analyzed in terms of PL, ML, FF, ER, and PE.

2.1 *Case 1: short narrow*

For case 1, a short narrow shading pattern is created with the shaded module which receives three different irradiations of 100, 200, and 300 W/m^2, and the unshaded module receives irradiation of 1000 W/m^2, as depicted in Fig. 11.1. These types of shading conditions are applied to the 4 $\times$ 4 PV array configurations constructed for the short narrow condition.

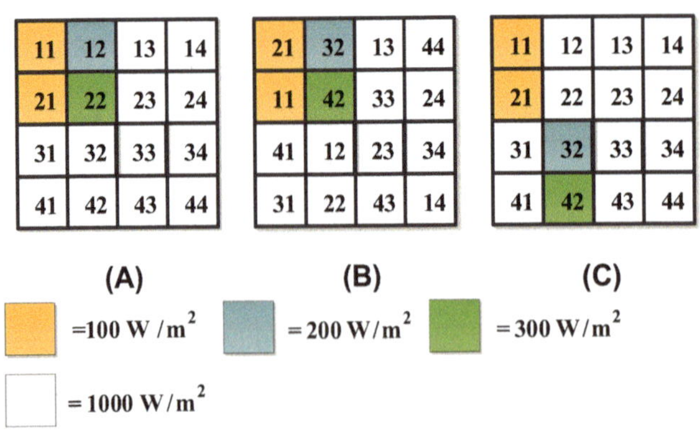

Figure 11.1 Partial shading pattern in case 1 short narrow: (A) TCT configuration; (B) JS configuration; (C) partial shading dispersion with JS puzzle arrangement.

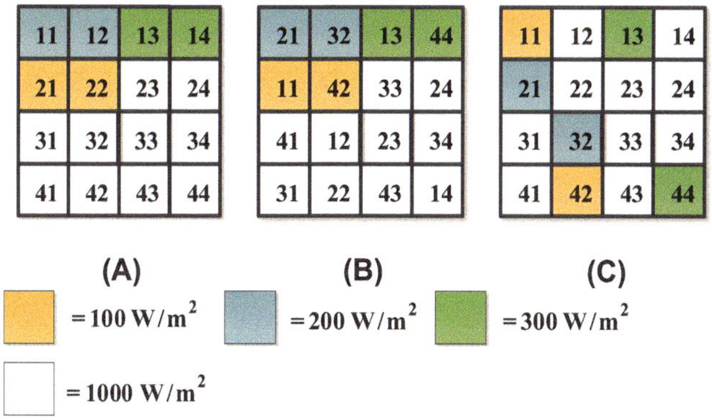

Figure 11.2 Partial shading pattern in case 2 short wide: (A) TCT configuration; (B) JS configuration; (C) partial shading dispersion with JS puzzle arrangement.

2.2 Case 2: short wide

For case 2, a short wide shading pattern is created with the shaded module which receives three different irradiations of 100, 200, and 300 W/m^2, and the unshaded module receives irradiation of 1000 W/m^2, as depicted in Fig. 11.2. From Fig. 11.2, it is clear that the impact of shading on the PV array is increased, which results in a reduction of maximum power.

2.3 Case 3: long narrow

For case 3, a long narrow shading pattern is created with the shaded module which receives three different irradiations of 100, 200, and 400 W/m^2, and the unshaded module receives irradiation of 1000 W/m^2, as depicted in Fig. 11.3. From Fig. 11.3,

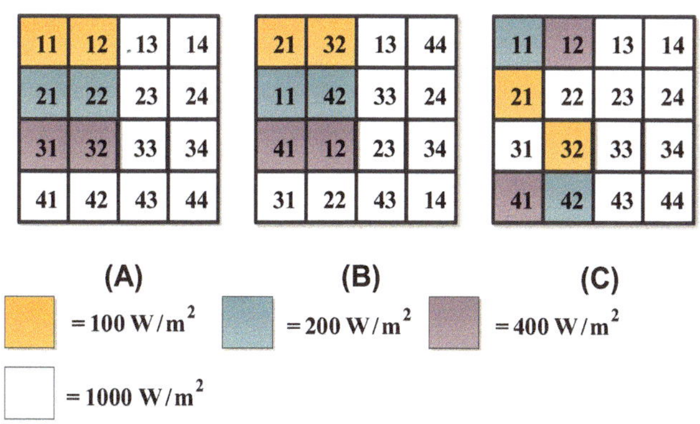

Figure 11.3 Partial shading pattern in case 3 long narrow: (A) TCT configuration; (B) JS configuration; (C) partial shading dispersion with JS puzzle arrangement.

it is understood that the impact of shading on a greater number of rows is increased as compared to short narrow conditions.

2.4 Case 4: long wide

For case 4, a long wide shading pattern is created with the shaded module which receives five different irradiations of 100, 200, 300, 400, and 500 W/m^2 and the unshaded module receives irradiation of 1000 W/m^2, as depicted in Fig. 11.4. From Fig. 11.4, it is clear that the impact of shading is increased as compared to other shading conditions.

3. Performance parameter

3.1 % Power loss

The percentage PL is the ratio of GMPP at the standard test condition (STC) and PSC to the GMPP at STC.

$$\%PL = G_{MPP}\ \text{at}\ STC - G_{MPP}\ \text{at}\ PSC/G_{MPP}\ \text{at}\ STC \tag{11.1}$$

3.2 Mismatch loss

ML refers to the difference in power between STC and PSC.

$$ML = G_{MPP}\ \text{at}\ STC - G_{MPP}\ \text{at}\ PSC \tag{11.2}$$

3.3 Fill factor

The FF is the squareness of the I−V curve. In STC, the I−V characteristics curve is rectangular, and under PSC it deviates from its shape. As the FF is close to unity,

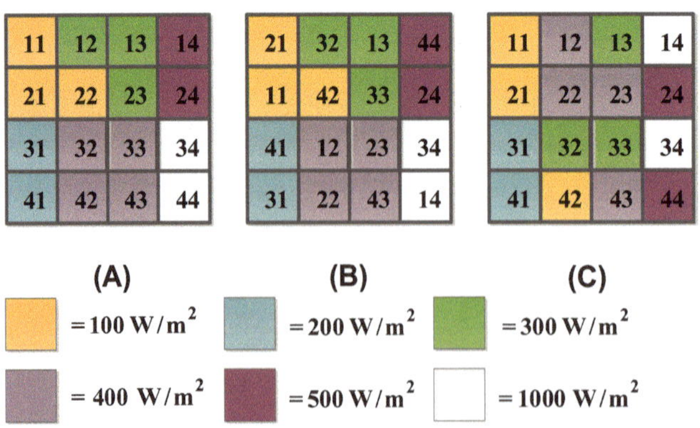

Figure 11.4 Partial shading pattern in case 4 long wide: (A) TCT configuration; (B) JS configuration; (C) partial shading dispersion with JS puzzle arrangement.

the PV array performance is higher. Hence, it is defined as the ratio of maximum power under PSC to the power generated.

$$FF = \frac{V_{mp} \times I_{mp} \text{ at PSC}}{V_{oc} \times I_{sc}} \tag{11.3}$$

3.4 Execution ratio

The ER is defined as the ratio of global maximum power point (GMPP) at PSC to GMPP at STC.

$$\text{ER}(\%) = \frac{\text{G}_{\text{MPP}} \text{ at PSC}}{\text{G}_{\text{MPP}} \text{ at STC}} \tag{11.4}$$

3.5 % Performance enhancement ratio compared to the conventional configuration

% PE is defined as the ratio of the difference in power between GMPP at STC and PSCs.

$$\%\text{PE} = \frac{\text{P}_{\text{GMPP}} \text{ (Proposed, KK, SS, OE, LS)} - \text{P}_{\text{GMPP}}\text{(TCT, SPTCT, BLTCT, HCTCT)}}{\text{P}_{\text{GMPP}} \text{ (Proposed, KK, SS, OE, LS)}} \tag{11.5}$$

4. Results and discussion

4.1 Case 1 short narrow

The performance analysis of the JS puzzle pattern with short narrow shading condition in comparison with the conventional [total cross-tied (TCT), series-parallel total cross-tied (SPTCT), bridge-link total cross-tied (BLTCT), and honeycomb total cross-tied (HCTCT)] configurations and rearranged hybrid configurations of KK, SS, OE, and LS puzzle pattern is made. The result in the form of GMPP, PL, ML, FF, ER, and PE is presented in Table 11.1.

4.1.1 TCT configuration

Results obtained from the conventional total cross-tied (CTCT) and rearranged total cross-tied (RTCT) configurations for the various puzzle patterns with short narrow shading condition indicate that the JS puzzle pattern performs superior and generates GMPP of 2493 W (Fig. 11.5A). The GMPP of the KK puzzle pattern is very close to the JS puzzle pattern with a GMPP of 2490 W. The SS, LS, OE, and CTCT follow the performance by 2120, 2119, 2043, and 2005 W. In the TCT configuration, since the

Table 11.1 Simulation results for case 1.

Short narrow	Puzzle pattern	V_{mp} (V)	I_{mp} (A)	GMPP (W)	PL (%)	ML (W)	FF (%)	ER (%)	PE (%)
TCT	CTCT	18.27	109.66	2005	37.34	1195.00	46.84	62.66	NA
	KK	24.22	102.77	2490	22.20	710.33	58.16	77.80	19.49
	SS	19.19	110.45	2120	33.76	1080.34	49.52	66.24	5.43
	OE	18.23	112.07	2043	36.16	1157.26	47.72	63.84	1.87
	LS	19.24	110.17	2119	33.77	1080.70	49.51	66.23	5.41
	JS	23.99	103.92	2493	22.10	707.30	58.23	77.90	19.58
SPTCT	CSPTCT	18.09	109.66	1984	38.00	1216.02	46.35	62.00	NA
	KK	17.78	105.41	1874	41.43	1325.62	43.79	58.57	−5.85
	SS	18.25	107.03	1954	38.95	1246.43	45.64	61.05	−1.56
	OE	23.84	80.30	1914	40.19	1286.04	44.71	59.81	−3.66
	LS	18.91	107.53	2033	36.45	1166.53	47.50	63.55	2.43
	JS	23.93	103.92	2487	22.29	713.26	58.09	77.71	20.22
BLTCT	CBLTCT	18.15	109.66	1991	37.79	1209.29	46.50	62.21	NA
	KK	23.52	105.41	2479	22.53	720.90	57.91	77.47	19.70
	SS	18.66	109.66	2046	36.07	1154.10	47.79	63.93	2.70
	OE	18.51	109.44	2026	36.70	1174.27	47.32	63.30	1.73
	LS	19.07	107.53	2051	35.91	1149.06	47.91	64.09	2.94
	JS	23.97	103.92	2491	22.15	708.81	58.19	77.85	20.09
HCTCT	CHCTCT	18.15	109.66	1991	37.79	1209.29	46.50	62.21	NA
	KK	23.54	105.41	2482	22.45	718.32	57.97	77.55	19.78
	SS	19.16	110.28	2113	33.97	1087.13	49.36	66.03	5.78
	OE	18.54	109.44	2029	36.59	1170.73	47.40	63.41	1.90
	LS	18.66	110.17	2056	35.76	1144.26	48.02	64.24	3.16
	JS	23.99	103.92	2493	22.10	707.30	58.23	77.90	20.14

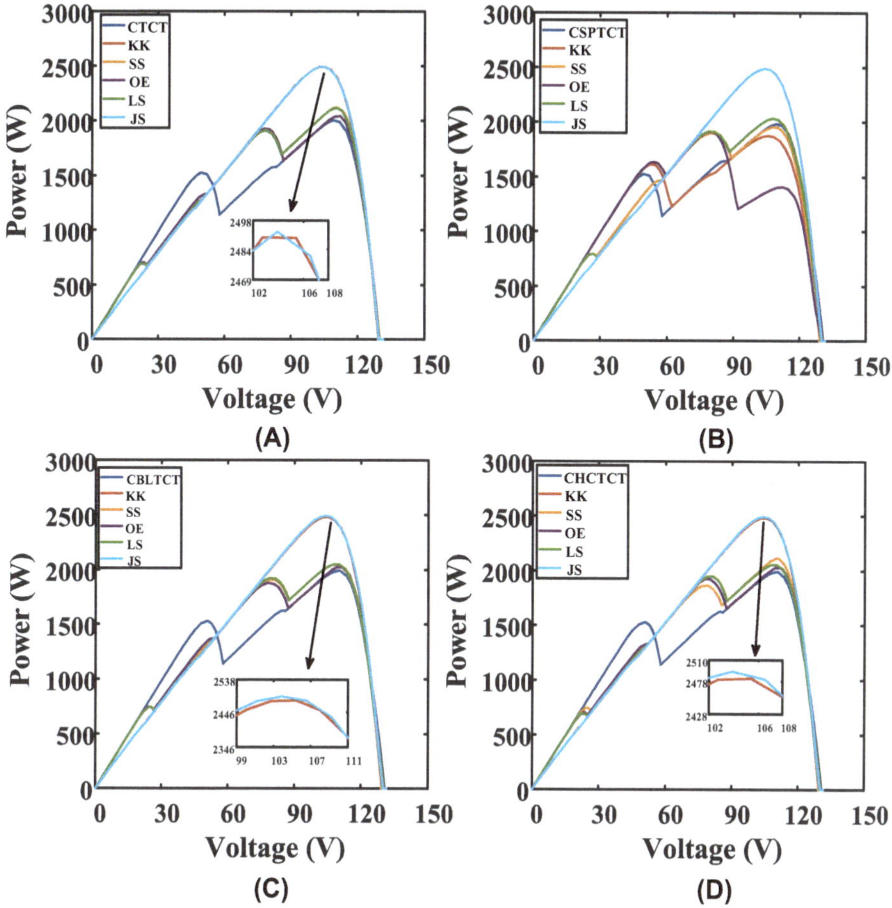

Figure 11.5 P–V characteristics: (A) TCT configuration; (B) SPTCT configuration; (C) BLTCT configuration; (D) SPTCT configuration for case 1.

ties are connected across the junction, the obtained GMPP is based on the effectiveness of the puzzle pattern. Hence, the JS puzzle pattern of TCT configuration has generated the relatively highest power output under short narrow conditions.

From Table 11.1, it is observed that the JS puzzle pattern has an ER ratio of 77.90%, followed by the KK, SS, LS, OE, and CTCT with 77.80%, 66.24%, 66.23%, 63.84%, and 62.66%. Also, the JS puzzle pattern has a PE ratio of 19.58% in comparison with the CTCT, followed by the KK, SS, LS, and OE with 19.49%, 5.43%, 5.41%, and 1.87%, respectively.

4.1.2 SPTCT configuration

The results obtained from the conventional series-parallel total-cross-tied (CSPTCT) and rearranged series-parallel total cross-tied (RSPTCT) configurations for the various puzzle patterns with short narrow shading conditions indicate that the JS puzzle pattern performs superior and generates GMPP of 2487 W (Fig. 11.5B). The LS, CSPTCT, SS, OE, and KK puzzle patterns follow the performance of the JS puzzle pattern with 2033, 1984, 1954, 1914, and 1874 W. Thus, it is noticed that the performance of the JS puzzle pattern generates 454 W more output power than the LS pattern. It is noticed that the JS puzzle pattern of the SPTCT configuration has generated the relatively highest power output under short narrow conditions since the SPTCT configuration has a higher number of series-connected modules compared to other configurations. At the time of rearranging, the modules according to the various puzzle patterns and the shaded modules are connected in series, thereby reducing the output power of the PV array.

From Table 11.1, it is observed that the JS puzzle pattern has an ER ratio of 77.71%, followed by the LS, CSPTCT, SS, OE, and KK puzzle patterns with 63.55%, 62.00%, 61.05%, 59.81%, and 58.57%. Also, the JS puzzle pattern has a PE ratio of 20.22% in comparison with CSPTCT, followed by the LS with 2.43%. The SS, OE, and KK have reductions in the PE ratio of −1.56%, −3.66%, and −5.85%, in comparison with the CSPTCT.

4.1.3 BLTCT configuration

The results obtained from the conventional bridge-link total-cross-tied (CBLTCT) and rearranged bridge-link total cross-tied (RBLTCT) configurations for the various puzzle patterns with short narrow shading conditions indicate that the JS puzzle pattern performs superior and generates a GMPP of 2491 W (Fig. 11.5C). The GMPP of the KK is very close to the JS puzzle pattern with a GMPP of 2479 W. The LS, SS, OE, and CBLTCT patterns follow the performance with 2051, 2046, 2026, and 1991 W. Hence, it is noticed that the JS puzzle pattern of the BLTCT configuration generates the relatively highest power output under short narrow conditions. Here, the number of parallel-connected modules in a row is reduced compared to the TCT configuration, and hence the maximum power is reduced.

From Table 11.1, it is observed that the JS puzzle pattern has an ER ratio of 77.85%, followed by the KK, LS, SS, OE, and CBLTCT patterns with 77.47%, 64.09%, 63.93%, 63.30%, and 62.21%. Also, the JS puzzle pattern has a PE ratio of 20.09% in comparison with the C-BLTCT, followed by the KK, LS, SS, OE, and CBLTCT patterns with 19.70%, 2.94%, 2.70%, and 1.73%, respectively.

4.1.4 HCTCT configuration

The results obtained from the conventional honeycomb total-cross-tied (CHCTCT) and rearranged bridge-link total cross-tied (RHCTCT) configurations for the various puzzle patterns with short narrow shading condition indicate that the JS puzzle pattern performs superior and generates a GMPP of 2493 W (Fig. 11.5D). The GMPP of the KK is very close to the JS puzzle pattern with a GMPP of 2482 W. The SS, LS, OE, and CHCTCT patterns follow this performance with 2113, 2056, 2029, and 1991 W. Thus, it is noticed that the JS puzzle pattern of the HCTCT configuration has generated the relatively highest power output under short narrow condition. Here, the number of parallel-connected modules in a row is high compared to the BLTCT, and hence the obtained GMPP is improved.

From Table 11.1, it is observed that the JS puzzle pattern has an ER ratio of 77.90%, followed by the KK, SS, LS and OE, and CHCTCT with 77.55%, 66.03%, 64.24%, 63.41%, and 62.21%. Also, the JS puzzle pattern has a PE ratio of 20.14% in comparison with the CHCTCT, followed by the KK, SS, LS, and OE with 19.78%, 5.78%, 3.16%, 1.90%, and 1.73% respectively.

4.2 Case 2 short wide

4.2.1 TCT configuration

The results obtained from the CTCT and RTCT configurations for the various puzzle patterns with short wide shading conditions indicate that the OE puzzle pattern performs superior and generates a GMPP of 2044 W (Fig. 11.6A). The GMPPs of the

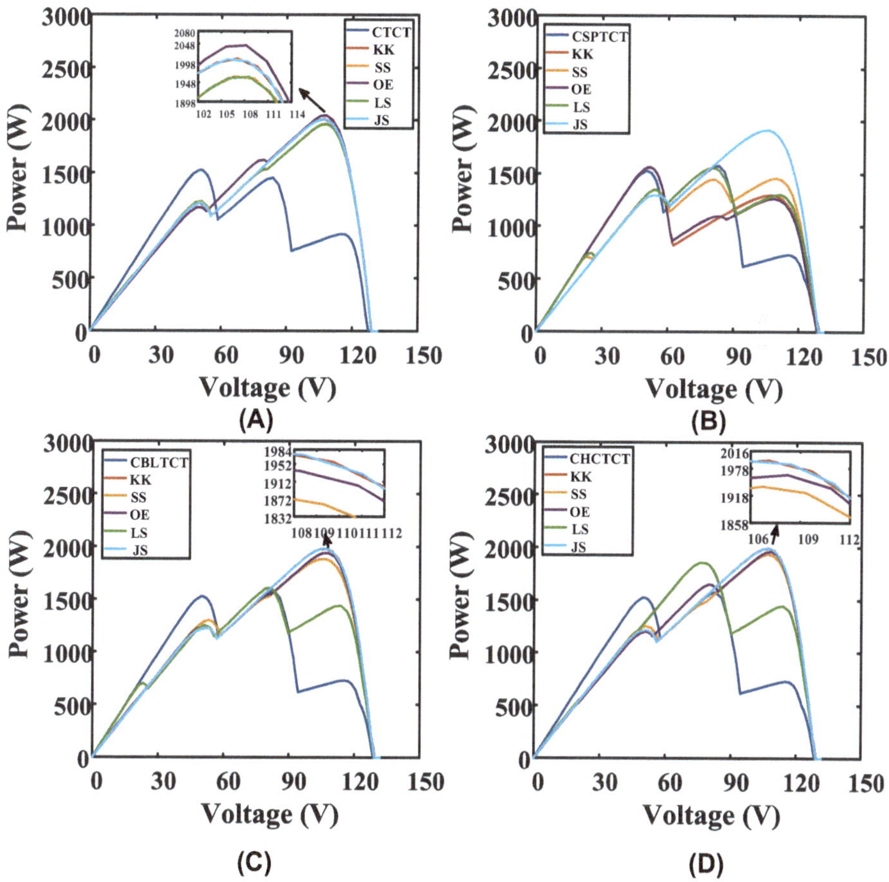

Figure 11.6 P–V characteristics: (A) TCT configuration; (B) SPTCT configuration; (C) BLTCT configuration; (D) SPTCT configuration for case 2.

KK and JS puzzle patterns are close to the OE pattern with GMPPs of 2010 and 2008 W. The SS, LS, and CTCT patterns follow this performance with 1963, 1963, and 1527 W. Thus, it is noticed that the OE puzzle pattern of the TCT configuration has generated the relatively highest power output under short wide conditions.

From Table 11.2, it is observed that the OE pattern has an ER ratio of 63.88%, followed by the KK, JS, SS, LS, and CTCT patterns with 62.81%, 62.76%, 61.35%, 61.33%, and 47.74%. Also, the OE puzzle pattern has a PE ratio of 25.27% in comparison with the CTCT, followed by the KK, JS, SS, and LS with 24.00%, 23.93%, 22.19%, and 22.16%, respectively.

4.2.2 SPTCT configuration

The results obtained from the CSPTCT and RSPTCT configurations for the various puzzle patterns with short wide shading conditions indicate that the JS puzzle pattern performs superior and generates a GMPP of 1915 W (Fig. 11.6B). The CSPTCT, OE, KK, LS, and SS follow this performance with 1579, 1563, 1561, 1557, and 1459 W. Thus, it is noticed that the JS puzzle pattern of the SPTCT configuration has generated the relatively highest power output short wide condition.

From Table 11.2, it is observed that the proposed puzzle pattern has an ER ratio of 59.83%, followed by the CSPTCT, OE, KK, LS, and SS with 49.33%, 48.83%, 48.80%, 48.67%, and 45.58%. Also, the proposed puzzle pattern has a PE ratio of 17.55% in comparison with the conventional configuration, followed by reductions in performance of the OE, KK, LS, and SS by −1.02%, −1.10%, −1.37%, and −8.23%, respectively.

4.2.3 BLTCT configuration

The results obtained from the CBLTCT and RBLTCT configurations for the various puzzle patterns with short wide shading conditions indicate that the JS puzzle pattern performs superior and generates a GMPP of 1982 W (Fig. 11.6C). The GMPP of the KK is very close to the JS puzzle pattern with a GMPP of 1981 W. The OE, SS, LS, and CBLTCT follow this performance with 1937, 1882, 1607, and 1579 W. Thus, it is noticed that the proposed and the KK puzzle patterns of the BLTCT configuration have generated almost identical maximum output powers under short wide condition.

From Table 11.2, it is observed that the proposed puzzle pattern has an ER ratio of 61.95%, followed by the KK, OE, SS, LS, and CBLTCT configurations with 61.91%, 60.55%, 58.81%, 50.23%, and 49.33%. Also, the JS puzzle pattern has a PE ratio of 20.37% in comparison with the CBLTCT, followed by reductions in performance of the KK, OE, SS, and LS by 20.32%, 18.52%, 16.12%, and 1.80%, respectively.

4.2.4 HCTCT configuration

The results obtained from the CHCTCT and RHCTCT configurations for the various puzzle patterns with short wide shading conditions indicate that the proposed and the KK patterns perform superior and generate a GMPP of 1995 W (Fig. 11.6D). The OE, SS, LS, and CHCTCT follow this performance with 1964, 1939, 1857, and 1579 W. Thus, it is noticed that the proposed and the KK puzzle patterns of the HCTCT

Table 11.2 Simulation results for case 2.

Short wide	Puzzle pattern	V_{mp} (V)	I_{mp} (A)	GMPP (W)	PL (%)	ML (W)	FF (%)	ER (%)	PE (%)
TCT	CTCT	30.38	50.28	1527	52.26	1672.32	35.69	47.74	NA
	KK	18.77	107.08	2010	37.19	1190.01	46.95	62.81	24.00
	SS	18.40	106.69	1963	38.65	1236.69	45.86	61.35	22.19
	OE	18.89	108.23	2044	36.12	1155.77	47.75	63.88	25.27
	LS	18.06	108.64	1963	38.67	1237.49	45.84	61.33	22.16
	JS	18.98	105.80	2008	37.24	1191.82	46.91	62.76	23.93
SPTCT	CSPTCT	18.97	83.22	1579	50.67	1621.40	36.88	49.33	NA
	KK	30.00	52.04	1561	51.20	1638.55	36.48	48.80	−1.10
	SS	13.34	109.32	1459	54.42	1741.47	34.07	45.58	−8.23
	OE	30.23	51.69	1563	51.17	1637.39	36.50	48.83	−1.02
	LS	19.39	80.30	1557	51.33	1642.71	36.38	48.67	−1.37
	JS	18.10	105.80	1915	40.17	1285.33	44.73	59.83	17.55
BLTCT	CBLTCT	18.97	83.22	1579	50.67	1621.40	36.88	49.33	NA
	KK	18.50	107.08	1981	38.09	1218.73	46.28	61.91	20.32
	SS	17.64	106.69	1882	41.19	1318.01	43.96	58.81	16.12
	OE	18.35	105.60	1937	39.45	1262.53	45.26	60.55	18.52
	LS	19.79	81.21	1607	49.77	1592.51	37.55	50.23	1.80
	JS	18.74	105.80	1982	38.05	1217.68	46.31	61.95	20.37
HCTCT	CHCTCT	18.97	83.22	1579	50.67	1621.40	36.88	49.33	NA
	KK	18.63	107.08	1995	37.65	1204.69	46.61	62.35	20.88
	SS	18.18	106.69	1939	39.40	1260.85	45.30	60.60	18.59
	OE	18.15	108.23	1964	38.62	1235.69	45.89	61.38	19.64
	LS	24.43	76.01	1857	41.98	1343.25	43.37	58.02	14.98
	JS	18.86	105.80	1995	37.65	1204.85	46.61	62.35	20.88

configuration have generated almost identical maximum output powers under short wide condition.

From Table 11.2, it is observed that the JS and the KK puzzle patterns have an ER ratio of 62.35%, followed by the OE, SS, LS, and CHCTCT with 61.38%, 60.60%, 58.02%, and 49.33%. Also, the JS and the KK puzzle patterns have a PE ratio of 20.88% in comparison with the CHCTCT, followed by reductions in performance of the OE, SS, and LS with 19.64%, 18.59%, and 14.98% respectively.

4.3 Case 3 long narrow

4.3.1 TCT configuration

The results obtained from the CTCT and RTCT configurations for the various puzzle patterns with long narrow shading conditions indicate that the JS puzzle pattern

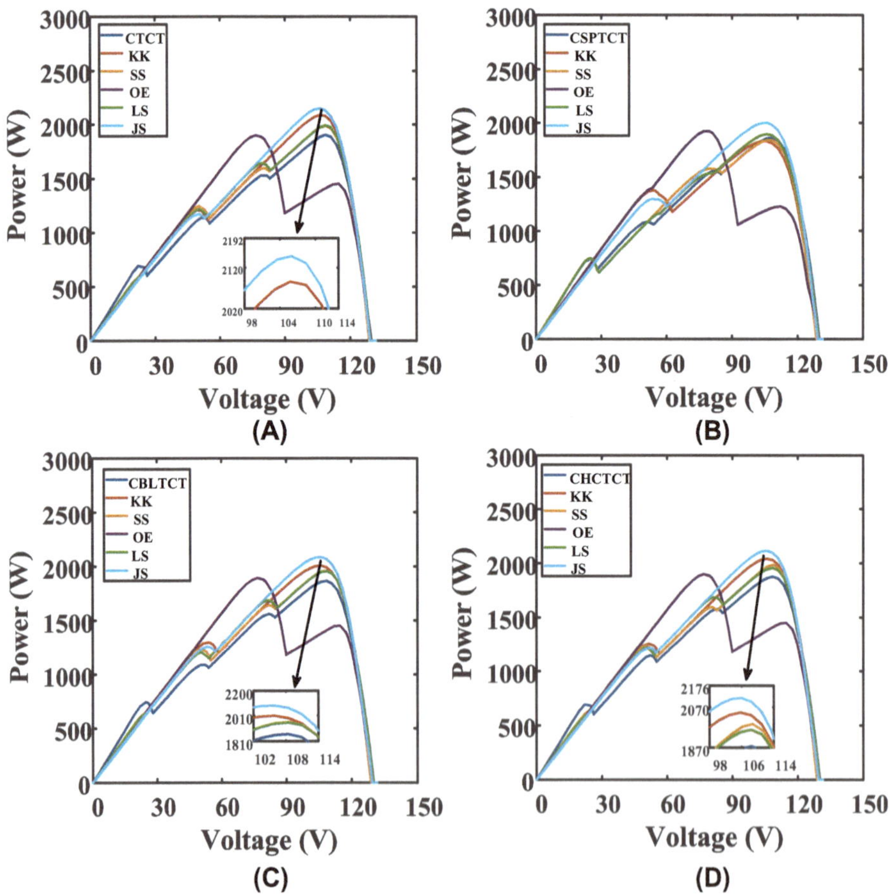

Figure 11.7 P–V characteristics: (A) TCT configuration; (B) SPTCT configuration; (C) BLTCT configuration; (D) SPTCT configuration for case 3.

performs superior and generates a GMPP of 2148 W (Fig. 11.7A). The KK, LS, SS, CTCT, and OE patterns follow this performance with 2085, 1993, 1900, 1985, and 1903 W. Thus, it is noticed that the JS puzzle pattern of the SPTCT configuration has generated a relatively highest output power under long narrow condition.

From Table 11.3, it is observed that the JS puzzle pattern has an ER ratio of 67.14%, followed by the KK, LS, SS, CTCT, and OE with 65.16%, 62.68%, 62.03%, 59.47%, and 59.38%. Also, the JS puzzle pattern has a PE ratio of 11.41% in comparison with the CTCT, followed by reductions in performance of the KK, LS, SS, and OE by 8.73%, 4.51%, 4.12%, and −0.16%, respectively.

4.3.2 SPTCT configuration

The results obtained from the CSPTCT and RSPTCT configurations for the various puzzle patterns with long narrow shading conditions indicate that the JS puzzle pattern performs superior and generates a GMPP of 2000 W (Fig. 11.7B). The OE, LS, CSPTCT, SS, and KK patterns follow this performance with 1924, 1896, 1865, 1844, and 1832 W. Thus, it is noticed that the JS puzzle pattern of the SPTCT configuration has generated a relatively highest output power under long narrow condition.

From Table 11.3, it is observed that the proposed puzzle pattern has an ER ratio of 62.50%, followed by the OE, LS, CSPTCT, SS, and KK patterns with 60.14%, 59.26%, 58.27%, 57.62%, and 57.26%. Also, the JS puzzle pattern has a PE ratio of 6.77% in comparison with the CSPTCT, followed by reductions in performance of the OE, LS, SS, and KK patterns of 3.10%, 1.67%, −1.13%, and −1.76%, respectively.

4.3.3 BLTCT configuration

The results obtained from the CBLTCT and RBLTCT configurations for the various puzzle patterns with long narrow shading conditions indicate that the JS puzzle pattern performs superior and generates a GMPP of 2086 W (Fig. 11.7C). The GMPP of the KK is very close to the JS puzzle pattern with a GMPP of 2005 W. The SS, LS, OE, and CBLTCT patterns follow this performance with 1958, 1955, 1892, and 1865 W. Thus, it is noticed that the JS puzzle pattern of the BLTCT configuration has generated the relatively highest output power under long narrow conditions.

From Table 11.3, it is observed that the JS puzzle pattern has an ER ratio of 62.67%, followed by the KK, SS, LS, OE, and CBLTCT with 62.67%, 61.18%, 61.11%, 59.14%, and 58.27%. Also, the JS puzzle pattern has a PE ratio of 10.63% in comparison with the CBLTCT, followed by reductions in performance of the KK, SS, LS, and OE patterns by 7.02%, 4.76%, 4.64%, and 1.47%, respectively.

4.3.4 HCTCT configuration

The results obtained from the CHCTCT and RHCTCT configurations for the various puzzle patterns with long narrow shading conditions indicate that the JS puzzle pattern performs superior and generates a GMPP of 2114 W (Fig. 11.7D). The GMPP of the KK is very close to the JS puzzle pattern with a GMPP of 2042 W. The SS, LS, OE,

Table 11.3 Simulation results for case 3.

Long narrow	Puzzle pattern	V_{mp} (V)	I_{mp} (A)	GMPP (W)	PL (%)	ML (W)	FF (%)	ER (%)	PE (%)
TCT	CTCT	17.56	108.36	1903	40.53	1296.84	44.46	59.47	NA
	KK	19.72	105.77	2085	34.84	1114.72	48.71	65.16	8.73
	SS	18.24	108.81	1985	37.97	1215.00	46.37	62.03	4.12
	OE	24.72	76.88	1900	40.62	1299.92	44.39	59.38	−0.16
	LS	18.39	108.36	1993	37.72	1207.02	46.56	62.28	4.51
	JS	20.28	105.94	2148	32.86	1051.65	50.19	67.14	11.41
SPTCT	CSPTCT	17.21	108.36	1865	41.73	1335.31	43.56	58.27	NA
	KK	17.32	105.77	1832	42.74	1367.60	42.81	57.26	−1.76
	SS	17.37	106.18	1844	42.38	1356.11	43.07	57.62	−1.13
	OE	24.20	79.51	1924	39.86	1275.63	44.95	60.14	3.10
	LS	17.94	105.73	1896	40.74	1303.67	44.30	59.26	1.67
	JS	18.88	105.94	2000	37.50	1199.87	46.72	62.50	6.77
BLTCT	CBLTCT	17.21	108.36	1865	41.73	1335.31	43.56	58.27	NA
	KK	18.96	105.77	2005	37.33	1194.60	46.85	62.67	7.02
	SS	17.99	108.81	1958	38.82	1242.21	45.73	61.18	4.76
	OE	24.62	76.88	1892	40.86	1307.54	44.21	59.14	1.47
	LS	18.05	108.36	1955	38.89	1244.51	45.68	61.11	4.64
	JS	19.69	105.94	2086	34.80	1113.61	48.74	65.20	10.63
HCTCT	CHCTCT	17.31	108.36	1876	41.37	1323.89	43.83	58.63	NA
	KK	19.31	105.77	2042	36.18	1157.63	47.71	63.82	8.14
	SS	18.24	108.81	1985	37.98	1215.47	46.36	62.02	5.46
	OE	24.72	76.88	1900	40.62	1299.92	44.39	59.38	1.26
	LS	18.06	108.36	1957	38.86	1243.49	45.70	61.14	4.11
	JS	19.96	105.94	2114	33.92	1085.57	49.39	66.08	11.27

and CHCTCT follow this performance with 1985, 1957, 1900, and 1876 W. Thus, it is noticed that the performance of the JS puzzle pattern generates 72W more output power than the KK pattern. Thus, it is noticed that the JS puzzle pattern of the HCTCT configuration has generated the relatively highest power output power under long narrow conditions.

From Table 11.3, it is observed that the JS puzzle pattern has an ER ratio of 66.08%, followed by the KK, SS, LS, OE, and CHCTCT with 63.82%, 62.02%, 61.14%, 59.38%, and 58.63%. Also, the JS puzzle pattern has a PE ratio of 11.27% in comparison with the CHCTCT, followed by reductions in performance of the KK, SS, LS, OE, and CHCTCT by 8.14%, 5.46%, 4.11%, and 1.26% respectively.

4.4 Case 4 long wide

4.4.1 TCT configuration

The results obtained from the CTCT and RTCT configurations for the various puzzle patterns with long wide shading conditions indicate that the KK, SS, LS, and JS puzzle patterns perform superior and generate a GMPP of 1390 W (Fig. 11.8A). The CTCT and OE patterns follow this performance with 1306 and 980 W.

From Table 11.4, it is observed that the KK, SS, LS, and JS puzzle patterns have an ER ratio of 43.43% followed by the CTCT and OE with 40.81% and 30.64%. Also, the KK, SS, LS, and JS puzzle patterns have a PE ratio of 6.03% in comparison with the CTCT, followed by a reduction in performance of the OE pattern by −33.20%, respectively.

4.4.2 SPTCT configuration

The results obtained from the CSPTCT and RSPTCT configurations for the various puzzle patterns with long wide shading condition indicate that the LS and JS puzzle patterns perform superior and generate a GMPP of 1349 W (Fig. 11.8B). The KK, SS, CSPTCT, and OE patterns follow this performance with 1339, 1339, 1305, and 985 W. Thus, it is noticed that LS and JS puzzle patterns of the SPTCT configuration have generated the relatively highest output power under long wide condition.

From Table 11.4, it is observed that the LS and JS puzzle patterns have an ER ratio of 42.17%, followed by the KK, SS, CSPTCT, and OE patterns by 41.85%, 41.85%, 40.78%, and 30.79%. Also, the LS and JS puzzle patterns have a PE ratio of 3.28% in comparison with the CSPTCT, followed by reductions in performance of the KK, SS, and OE patterns by 2.54%, 2.54%, and −32.45%, respectively.

4.4.3 BLTCT configuration

The results obtained from the CBLTCT and RBLTCT configurations for the various puzzle patterns with long wide shading conditions indicate that the KK and JS puzzle patterns perform superior and generate a GMPP of 1389 W (Fig. 11.8C). The SS, LS, CBLTCT, and OE patterns follow this performance with 1361, 1361, 1305, and

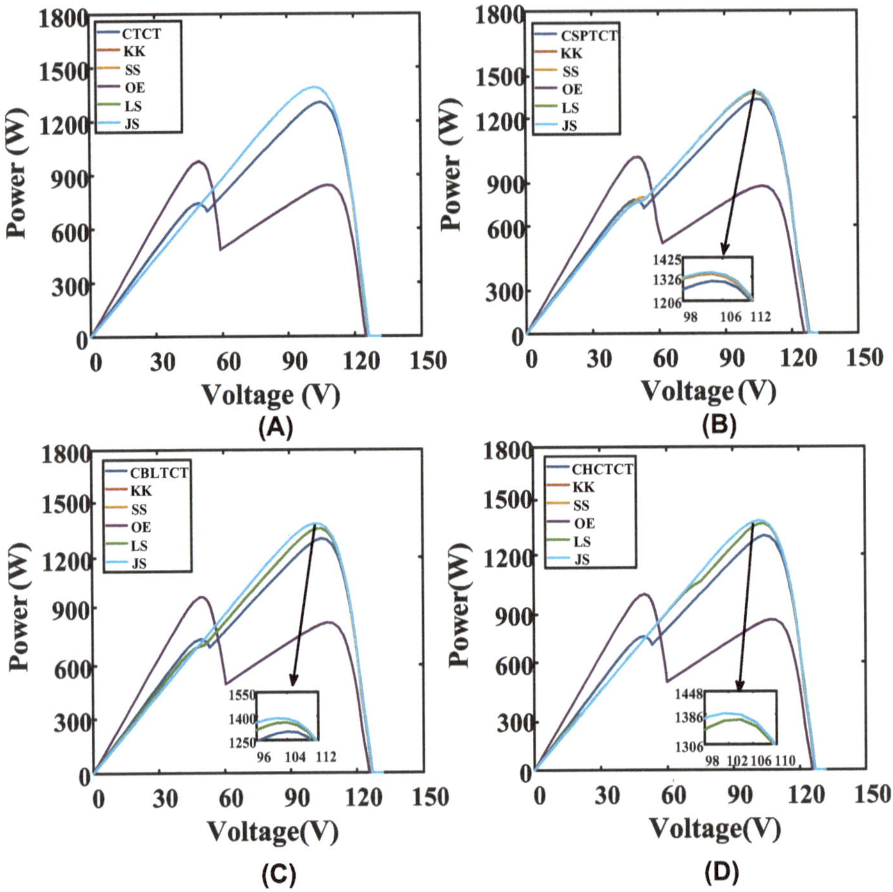

Figure 11.8 P–V characteristics: (A) TCT configuration; (B) SPTCT configuration; (C) BLTCT configuration; (D) SPTCT configuration for case 4.

980 W. Thus, it is noticed that the KK and JS puzzle patterns of the BLTCT configuration have generated the relatively highest output power under long wide conditions.

From Table 11.4, it is observed that the KK and JS puzzle patterns have an ER ratio of 43.41%, followed by the SS, LS, CBLTCT, and OE with 42.53%, 42.53%, 40.79%, and 30.64%. Also, the KK and JS puzzle patterns have a PE ratio of 6.04% in comparison with the CBLTCT, followed by a reduction in performance of the LS, SS, and OE patterns by 4.09%, 4.28%, and −33.14%, respectively.

4.4.4 HCTCT configuration

The results obtained from the CHCTCT and RHCTCT configurations for the various puzzle patterns with long wide shading conditions indicate that the KK and JS puzzle patterns perform superior and generate a GMPP of 1389 W (Fig. 11.8D). The SS, LS,

Table 11.4 Simulation results for case 4.

Long wide	Puzzle pattern	V_{mp} (V)	I_{mp} (A)	GMPP (W)	PL (%)	ML (W)	FF (%)	ER (%)	PE (%)
TCT	CTCT	12.46	104.82	1306	59.19	1894.14	30.51	40.81	NA
	KK	13.60	102.19	1390	56.57	1810.35	32.46	43.43	6.03
	SS	13.60	102.19	1390	56.57	1810.35	32.46	43.43	6.03
	OE	19.75	49.65	980	69.36	2219.59	22.90	30.64	−33.20
	LS	13.60	102.19	1390	56.57	1810.35	32.46	43.43	6.03
	JS	13.60	102.19	1390	56.57	1810.35	32.46	43.43	6.03
SPTCT	CSPTCT	12.57	103.81	1305	59.22	1894.92	30.49	40.78	NA
	KK	12.90	103.81	1339	58.15	1860.85	31.28	41.85	2.54
	SS	12.90	103.81	1339	58.15	1860.85	31.28	41.85	2.54
	OE	19.30	51.06	985	69.21	2214.64	23.02	30.79	−32.45
	LS	13.00	103.81	1349	57.83	1850.62	31.52	42.17	3.28
	JS	13.00	13.00	1349	57.83	1850.62	31.52	42.17	3.28
BLTCT	CBLTCT	12.56	103.96	1305	59.21	1894.63	30.49	40.79	NA
	KK	13.71	101.33	1389	56.59	1810.76	32.45	43.41	6.04
	SS	13.09	103.96	1361	57.47	1839.08	31.79	42.53	4.08
	OE	19.68	49.82	980	69.36	2219.56	22.90	30.64	−33.14
	LS	13.09	103.96	1361	57.47	1839.01	31.79	42.53	4.09
	JS	13.71	101.33	1389	56.59	1810.76	32.45	43.41	6.04
HCTCT	CHCTCT	12.56	103.95	1305	59.21	1894.64	30.49	40.79	NA
	KK	13.71	101.32	1389	56.59	1810.77	32.45	43.41	6.04
	SS	13.19	103.95	1372	57.14	1828.35	32.04	42.86	4.83
	OE	19.66	49.87	980	69.36	2219.56	22.90	30.64	−33.14
	LS	13.19	103.95	1371	57.14	1828.53	32.04	42.86	4.82
	JS	13.71	101.32	1389	56.59	1810.76	32.45	43.41	6.04

CHCTCT, and OE patterns follow this performance with 1372, 1371, 1305, and 980 W. Thus, it is noticed that the KK and JS puzzle patterns of the HCTCT configuration have generated the relatively highest power output power under long wide condition.

From Table 11.4, it is observed that the KK and JS puzzle patterns have an ER ratio of 43.41%, followed by the SS, LS, CHCTCT, and OE patterns with 42.86%, 42.86%, 40.79%, and 30.64%. Also, the KK and JS puzzle patterns have a PE ratio of 6.04% in comparison with the CHCTCT, followed by reductions in performance of the SS, LS, and OE patterns by 4.83%, 4.82%, and −33.14% respectively.

5. Experimental verification

A detailed experimental setup has been conducted on the 4 × 4 PV array configurations to validate the effectiveness of the proposed JS puzzle arrangements. The specification of the PV modules considered for the experimental setup is represented in Table 11.5. Fig. 11.9 shows the experimental setup of a 4 × 4 PV array configuration. Each module

Table 11.5 Electrical characteristics of the PV module for experimental verification.

Electrical characteristics	Ratings
Maximum power (P_{max})	10 W
Open circuit voltage (V_{oc})	21.96 V
Voltage at maximum power (V_{mpp})	18.25 V
Short circuit current (I_{sc})	0.59 A
Current at maximum power (I_{mpp})	0.55 A
Number of cells connected in series (N_s)	36

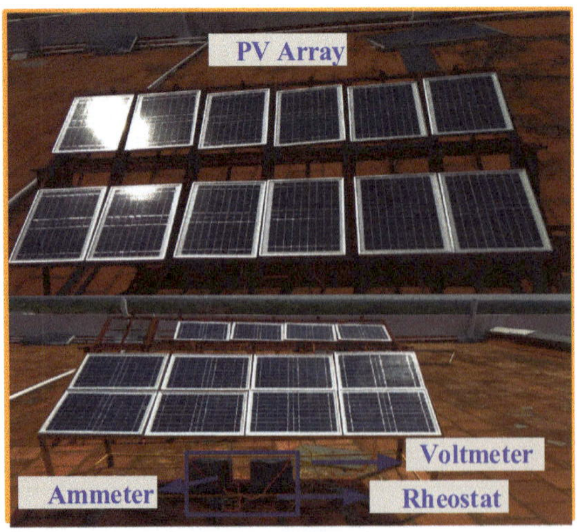

Figure 11.9 Experimental setup of a 4 × 4 PV array configuration.

of 10 W is grouped to form a 4 × 4 PV array. The PV array is connected to a variable resistance and the value of resistance is adjusted manually to achieve maximum power. The resistance load is set at a value of 33.33 Ω and the maximum power produced by the PV array is 145 W. The Experiment was conducted on a typical sunny day and the solar radiation was observed as 965 W/m^2 at the experimental spot.

The effects of partial shading are created on the PV array by covering the modules with cardboard. The PV array was tested under four PSCs such as one module, two-module, three modules, and four modules in a row being shaded. The TCT, SPTCT, BLTCT, and HCTCT configurations are tested for KK, SS, OE, LS, and JS under different PSCs. From the experimental result (Table 11.6), it is observed that for one module shading condition, SPTCT, BLTCT, and TCT produced maximum power of 116, 112, and 110 W, followed by the HCTCT configuration with the maximum power of 112 W for KK and SS and 110 W for CHCTCT, OE, LS, and JS puzzle patterns, respectively. Similarly, from the results of the other PSCs, it is observed that the

Table 11.6 Experimental results.

	Puzzle pattern	(One module shaded) power (W)	(Two modules shaded) power (W)	(Three modules shaded) power (W)	(Four modules shaded) power (W)
TCT	CTCT	110	110	110	110
	KK	110	117	116	115
	SS	110	117	116	115
	OE	110	117	79	79
	LS	110	117	116	115
	JS	110	117	116	115
	CSPTCT	116	112	115	110
	KK	116	114	84	76
SPTCT	SS	116	112	84	76
	OE	116	114	84	76
	LS	116	112	84	76
	JS	116	114	84	76
	CBLTCT	112	112	110	110
	KK	112	115	81	79
BLTCT	SS	112	117	115	115
	OE	112	115	81	79
	LS	112	117	115	115
	JS	112	115	81	79
HCTCT	CHCTCT	110	110	110	110
	KK	112	114	115	84
	SS	112	117	116	115
	OE	110	114	81	79
	LS	110	117	115	115
	JS	110	114	115	84

TCT configuration has produced a superior performance for almost all the puzzle patterns under the considered PSCs. The performance of the TCT is followed by the HCTCT, BLTCT, and SPTCT configurations.

6. Conclusion

In this chapter, the performance analysis of a 4 × 4 PV array with TCT and hybrid-rearranged configuration was carried out under PSCs. To validate the performance of the proposed JS puzzle pattern, the PV array was subjected to different partial shading patterns. The investigation was carried out in the MATLAB/SIMULINK environment with a bypass diode connected across each module. The performance of the JS puzzle pattern on a hybrid-rearranged configuration is compared to the existing puzzle patterns in terms of GMPP, PL, ML, FF, ER, and PE.

Overall, the simulation results obtained by the hybrid-rearranged configuration with the JS puzzle-based reconfiguration technique improved the maximum power by 14.01% under the PSCs. The performance of the JS puzzle-based reconfiguration technique is followed by the KK, SS, LS, and OE puzzle-based reconfiguration techniques of 10.00%, 5.73%, 5.29%, and −3.88%, respectively. From the experimental results, it is observed that the KK, SS, LS, and JS puzzle pattern-based TCT configuration produced superior results. Overall, the TCT configuration produced superior performance compared to other configurations under the PSCs. The performance of the TCT is followed by the HCTCT, BLTCT, and SPTCT configurations. The results show the effectiveness of the JS puzzle-based reconfiguration technique in dispersing the partial shading effects throughout the array compared to the existing puzzle-based reconfiguration techniques. The proposed technique eliminates the need for the complex algorithm, switching matrix, and sensors, which is economically viable for the small-scale PV array, whereas the large-scale PV array can be divided into a number subarray.

References

[1] S.P. Dhanup, J. Prasanth Ram, S. Veronika, K. Young-Jin, A new shade dispersion technique compatible for symmetrical and unsymmetrical photovoltaic (PV) arrays, Energy 225 (2021) 120241.

[2] M. Etarhouni, B. Chong, L. Zhang, A combined scheme for maximising the output power of a Photovoltaic array under partial shading conditions, Sustainable Energy Technologies and Assessments 50 (2022) 101878.

[3] V.K. Yadav, Y. Abhishek, Y. Ranjana, M. Aaradhya, H.W. Nadeem, G. Shubham, K.P. Rupendra, G. Santosh, A novel reconfiguration technique for improvement of PV reliability, Renewable Energy 182 (2022) 508−520.

[4] A.S. Yadav, R.K. Pachauri, Y.K. Chauhan, S. Choudhury, R. Singh, Performance enhancement of partially shaded PV array using novel shade dispersion effect on magic-square puzzle configuration, Solar Energy 144 (2017) 780−797.

[5] Y. Jiang, J. Xu, X. Leng, N. Eghbalian, A novel hybrid maximum power point tracking method based on improving the effectiveness of different configuration partial shadow, Sustainable Energy Technologies and Assessments 50 (2022) 101835.
[6] G. Sai Krishna, T. Moger, Enhancement of maximum power output through reconfiguration techniques under non-uniform irradiance conditions, Energy 187 (2019) 115917.
[7] I. Nasiruddin, K. Shahida, F.J. Mohd, R.C. Bansal, Shade diffusion of partial shaded PV array by using odd-even structure, Solar Energy 181 (2019) 519−529.
[8] R.K. Pachauri, A.S. Yadav, Y.K. Chauhan, A. Sharma, V. Kumar, Shade dispersion-based photovoltaic array configurations for performance enhancement under partial shading conditions, International Transactions on Electrical Energy Systems 28 (2018) 2556−2587.

Index

'*Note:* Page numbers followed by "f" indicate figures and "t" indicate tables.'

U

V

W

X

FSC
www.fsc.org
MIX
Papier | Fördert
gute Waldnutzung
FSC® C083411

Kolibri 360 GmbH
Ferdinand-Jühlke-Straße 7
99095 Erfurt, Deutschland
produktsicherheit@kolibri360.de